AF394361

How
Flowers Made
Our World

www.penguin.co.uk

How Flowers Made Our World

The Story *of* Nature's Revolutionaries

David George Haskell

WITH ILLUSTRATIONS BY LUCY SMITH

torva

TRANSWORLD PUBLISHERS

UK | USA | Canada | Ireland | Australia
India | New Zealand | South Africa

Transworld is part of the Penguin Random House group of companies
whose addresses can be found at global.penguinrandomhouse.com.

Penguin Random House UK, One Embassy Gardens,
8 Viaduct Gardens, London SW11 7BW

penguin.co.uk

First published in Great Britain in 2026 by Torva
an imprint of Transworld Publishers

001

Printed and bound in Great Britain by Clays Ltd, Elcograf S.p.A.

The authorized representative in the EEA is Penguin Random House Ireland,
Morrison Chambers, 32 Nassau Street, Dublin D02 YH68.

A CIP catalogue record for this book is available from the British Library.

ISBN:
9781911709985 (hb)
9781911709992 (tpb)

Penguin Random House is committed to a sustainable future
for our business, our readers and our planet. This book is made
from Forest Stewardship Council® certified paper.

Dedicated to my mother, Jean Haskell,

and my sister, Julia Grindley,

with love and gratitude

Contents

Preface ❧ *xi*

Magnolia ❧ *1*

Goatsbeard ❧ *23*

Orchid ❧ *45*

Grass ❧ *71*

Seagrass ❧ *103*

Rose ❧ *129*

Tea ❧ *159*

Pansy ❧ *183*

Speculative Futures ❧ *223*

Afterword:
On Floral Beauty and Joy ❧ 255

Acknowledgments ❧ 259

Supplement:
Invitations to Play with Flowers ❧ 263

Bibliography ❧ 283

Index ❧ 315

Preface

I'LL SERVE IT WITHOUT THE FLOWER," SAYS THE BAR-
tender as he stirs my cocktail.

"No! Bring it on, I love flowers." We share a smile.

He perches a single orchid bloom on the rim as he slides the glass
across the bar top. I feel a moment of guilt for the decapitated moth orchid, then raise the glass. My senses converge: purple glow of petals,
aromas of fruit and bitters, and the taste of gin. Like a happy bee, I sip.

"Yeah, guys will send the drink back if I garnish it with flowers or use
fancy stemware. They can't deal with it." As the bartender explains, the
couple next to me laugh and nod. She has a martini glass. He's drinking
beer straight out of the bottle. I look at the orchid perched on the rim of
my glass. Biologically, the flower, like most flowers, has both male and
female sex cells, making pollen and eggs. Yet, in this bar the orchid is
shoehorned into a symbol for a single human gender.

I'm in Knoxville, but masculine disdain for flowers reaches much farther than Bible Belt Tennessee. When a reporter for *The New York Times*,
Becky Hughes, visited bars in Manhattan, bartenders told her that many
men cannot abide drinks that, in their view, send the wrong message.
Flowers in the drink? A curvy glass? Nope. Too girlie.

This is just one of the ways that we box flowers into limited symbolic roles. In many parts of contemporary Western culture, as well as being seen as feminine, flowers are regarded as weak and merely ornamental. Flowers are pretty, but not strong or in charge. This way of thinking hides an essential truth. Flowers are world changers.

When flowers evolved, they upended and transformed the planet. They were late arrivals on the world stage, appearing about two hundred million years ago, long after the evolution of complex animals and other land plants. By one hundred million years ago they were the foundation of most habitats on land. Earth was forever changed by flowers. They created habitats like tropical rainforests and prairies, and catalyzed the evolution and diversification of bees, butterflies, birds, mammals, and many other animals. Today, 90 percent of all plant species on the planet are flowering plants. They provide food and habitat for much of life on Earth. Without them, humans would not have evolved, nor could we feed ourselves today. Flowering plants are some of life's great revolutionaries. They belong at the center of the story of how our world came to be.

Revolution is a word not often used by biologists. Evolution has wrought so many changes in three and a half billion years that those who study life's history have a low tolerance for exaggeration. There is no formal definition, but biological "revolutions" are events that set the world on entirely new courses, reworking the physical and chemical nature of Earth, and inventing new ways of life and new habitats. Revolutions remake the planet. The oxygen revolution was one, starting 2.4 billion years ago. Microbes and, later, algae and land plants suffused the water and air with oxygen, poisoning many ancient creatures but paving the way for the modern world. The explosion of animal diversity and the evolution of swimming animals 550 million and 400 million years ago were also revolutions. Flowering plants join this small pantheon. Flowers' evolution and eventual takeover of much of the world is called by biologists the "Cretaceous terrestrial revolution," for the Cretaceous period when they became abundant. This name from the technical literature is too modest, though. Flowering plants remade life in the oceans, as well as on land, and their revolutionary innovations continued long after their appearance and rise in the Cretaceous. We live on a floral planet.

Flowering plants caused an explosion of biodiversity and ecological productivity. Today, they are the foundations of most of the Earth's habitats. They transformed the climate, locally and globally. Not bad work for a group of creatures whose defining feature, the flower, usually lasts only a few days. The revolution brought by flowering plants was largely catalyzed by beauty and cooperation, with some genetic innovation and sensory illusion thrown in for good measure. Flowers are especially good at drawing together different species into unlikely, but highly productive, partnerships. By interweaving the lives of soil microbes and fungi with the aboveground world of pollinating insects and fruit-dispersing birds, for example, they ignited frenzies of evolution and created habitats whose ability to alchemize sunlight and air into life is unparalleled.

Today, flowering plants quite literally run most of the planet's ecology. We humans are especially indebted to them. Without grasses—recently evolved flowers that build sweeping expanses of food-filled prairies, savannas, and steppes—we'd still be small-brained apes swinging through the forest. Most of the calories we now eat come from grasses such as rice, maize, wheat, and sugarcane. Much of human agriculture focuses on keeping these grasses happy so that we, too, can thrive, a planetwide symbiosis between mammal and flower.

Symbiosis also manifests in our sensory entanglements with flowers, especially in gardens and through perfumery. There, the art and science of flower cultivation merge human desire and floral abundance. Amid the beauty that grows from our tending of flowers, though, brokenness and injustice sometimes emerge as we poison and disregard land, people, and pollinators in our quest for floral perfection.

Flowers also inspired intellectual revolutions. It was through the study of floral sexuality that modern biology, especially the study of classification and evolution, was born, strangely and tragically twinned with horrific ideas about race in humans. The enslavement of people was partly justified by European colonists with biological "classifications" that had their origins in the study of flowers but were then distorted into racist falsehoods when applied to human diversity.

Looking to the future, the resilience and creativity of flowers offer near-term visions of playful flowering weeds taking over a world upended by human destruction. This is no balm. The nimble flowering plants that come in our wake are impressively adaptable, but they are a shadow of the former diversity and abundance of floral life. Gazing further, tens of millions of years hence into a post-human world, flowers offer a surprising vision of beauty. In the past, places with poor soils, fire, and heat—the very conditions we are now imposing through our improvidence—have been crucibles from which emerge the world's most spectacular flowering plants. Such speculative visions underscore the need to protect the diversity of flowers today, for their vitality is the precondition for future resilience or renewal.

Flowers overturned the old order of Earth and built the world we live in. They will also shape its future. Yet, we seldom credit flowers with making our world. Send that idea back to the bar and take out the flower.

Our limited imagination about the astonishing creativity and productivity of flowers is part of a more general disregard for plants. Fewer than 1 percent of European Paleolithic cave paintings feature plants, and none represent flowers. A meat-heavy ice-age diet may partly explain this bias. In places like the Australian Kimberley and the American Southwest, flowers and leaves appear more frequently in rock art, although animals usually still dominate.

The bias continues today. We use plant metaphors to sneer at weak language. "Flowery" sentences use ornamentation to hide lack of substance. Real flowers are such successful communicators that whole dynasties of insects—butterflies and bees—evolved to read them. "Wooden" writing is poor, even though real wood is supple and strong. We admire the animal flesh in "muscular" language. Even the "flower power" of the 1960s denoted passivity, yet real flowers actively direct the animal world around them, and their prodigious growth builds new ecosystems. Consider, too, how we name the stages of life's history. Children's books, old biology textbooks, and modern nature documentaries mark the grand divisions using animals, the "Age of Dinosaurs" and then our present "Age of Mammals," even though the sunflower and orchid families each include more living species than all mammals and reptiles combined. Scientists use similar categories, dividing time into Paleozoic, Mesozoic, and Cenozoic—the "Ancient Animal," "Middle Animal," and "New Animal" eras. *Zoic* is from the Ancient Greek *zôion*, meaning animal, hence zoos and zoology, although the word sometimes encompasses all life.

Walk into most natural history museums and the story of life's past is told by animals. There, trilobite fossils, vertebrate bones, glass-eyed birds, and pinned butterflies tell us of life's grand narratives. In these zoocentric museum galleries, plants live in the background or underfoot. Flowers are usually barely mentioned. These animal-focused narratives of the past omit some of evolution's leading characters, the plants, especially flowering plants, that built and shaped the storyline. True, plants leave fewer fossils than animals, but the bigger problem is that we would rather see ferocity than flowers, *Tyrannosaurus* than tulips, and so there is little appetite for giving plants a leading role in re-creations of the story of life on Earth.

In Manhattan, for example, flowers are celebrated in museums devoted to aesthetic appreciation, but not in those whose focus is life's evolution. Art is floral, but science, apparently, is not. The American Museum of Natural History is packed with wonders and staffed by brilliant curators and scientists. Yet, as we enter a space whose name encompasses the history of all life, we are greeted by fighting dinosaurs standing on bare earth, not a shred of green to be seen, let alone a flower. Above them, a giant wall inscription proclaims Theodore Roosevelt's thoughts about "MANHOOD." These displays of violence and masculinity date to the early twentieth century, when the museum was expanded as a memorial to Roosevelt. To see flowers celebrated, we walk across Central Park to the Great Hall of the Metropolitan Museum of Art. There, gorgeous giant bouquets stand year-round in wall niches and at the central information desk, refreshed weekly, kept in perpetual bloom by a dedicated endowment.

Botanic gardens, strangely, sometimes reinforce the message that flowers are attractive but inconsequential. Many have reminders amid their floral displays about the practical utility of some plants (vanilla in the orchid room, cacao in the tropical glasshouse), but few send the message that flowers built most ecosystems and are now running the planet. One notable exception is the Royal Botanic Gardens, Kew, in London, where signs at the entrance proclaim, "If plant life stops, all life stops" and the most prominent link on the front page of the website invites us to "Discover the power of plants." *All life. Power.* These are words that break through cultural prejudice.

At Kew, the word *power* also carries a dark shadow of colonial history. The command gained by studying, transporting, and commercializing flowering plants—cotton, coffee, sugar, tea, indigo, rice, and many more—was a foundation of European colonialism. The scientific and commercial elite of eighteenth- and nineteenth-century Europe founded or expanded private and public botanic gardens to better explore and control the floral world. Unlike some of their descendants in American barrooms, these men had no problem with flowers. Instead, they built obscene empires with them. Ornament, science, and power were united in colonial botanic gardens, and the study of flowers was foundational to modern biology. Before this, kings, queens, and emperors sometimes used flowers as symbols of authority, associating beauty with dominion. The chrysanthemum is the imperial symbol of Japan, and roses and lilies signified royalty in Britain and France for centuries. Flowers, then, have not always and in all places been dismissed as, well, shrinking violets or wallflowers.

In this book, I aim to put flowers back where they belong, at the center of the story of how our world came to be. I also want to reclaim flowers for all human genders. I will succeed if you look on flowers with awed gratitude. If you order a bloom-garnished drink and tell florid stories to your bar companions, so much the better.

Luckily for us, our ancestors have done the introductory work. They bequeathed to us senses that purr in the presence of flowers. Eons of natural selection gave us bodies that respond to flowers with a rewarding shot of pleasure. Etched in our genes is old wisdom: Enjoy and be curious about the many pleasures of flowers. Hey, our forebears tell us, lean in and sniff. Enjoy the shapes and colors of petals. Give your lover a bloom. This book is an answer to these invitations, an exploration and a celebration of flowers. At the end of the book, I also offer some practical exercises with everyday flowers to deepen your curiosity and sensory delight.

A word about what I mean by *flowers*. The term has two related meanings, both of which I explore. Flowers are the sexy and parental parts of many plants, containing all that is needed to attract pollinators, unite egg and sperm, and then care for the developing embryos. Petals, pollen, ovaries, fruits. These are all flower. The other meaning is shorthand for "flowering plant," from roots to twig tips, the whole beings from which emerge the blooms.

Flowers—in both meanings of the word—made our world. To see how, let's start by following the scent of a blooming magnolia.

How
Flowers Made
Our World

Magnolia

Ancient flowers revolutionize the world

with beauty, bisexuality, and

motherhood

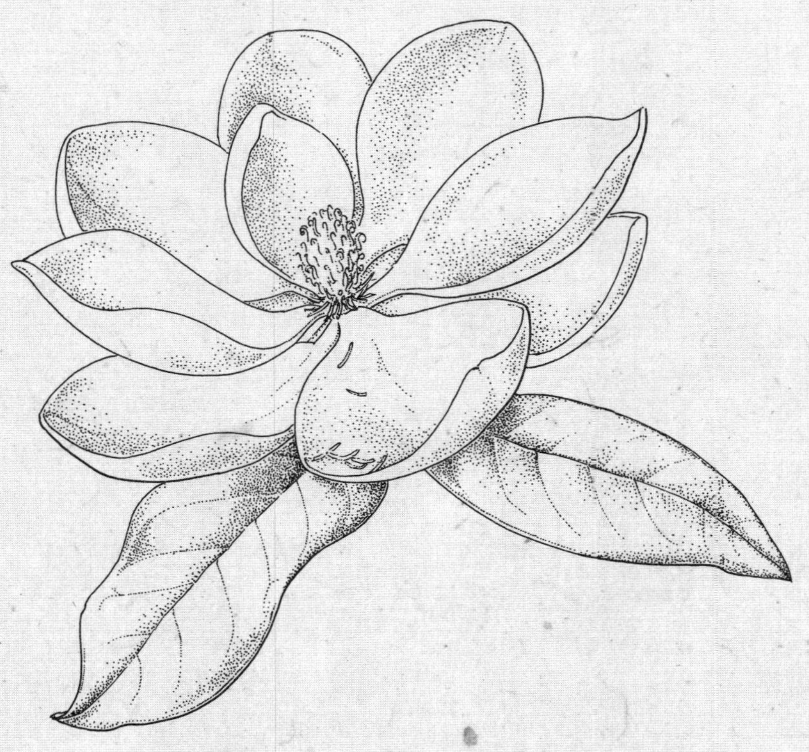

Ancient flowers revolutionize the world with beauty, bisexuality, and motherhood

A DELICIOUS AROMA ANNOUNCES SUMMER'S ARRIVAL in Atlanta. In front of office buildings in Midtown, my nose plunges into sweetened almond and tangerine, then resurfaces in a miasma of truck fumes. The scent brushes past me, too, on the scuffed lawns of city parks and along shady streets in older neighborhoods. To find the source of these delights, I look up. White flowers the size and color of porcelain dinner plates sprout from the branch tips of huge southern magnolia trees. A plume of scent streams from each bloom. These aromatic trails dance like ribbons in a breeze, so I can never know when one will warm me with its touch. Over the next month, this scant first bloom will grow. By June, a large tree can have one hundred or more flowers open at the same time.

The delights of magnolia flowers are portals into deep time. When *Tyrannosaurus* stomped through the subtropical forests of the late Cretaceous, magnolia-like plants had been thriving for fifty or more million years. If magnolias were rare animals, they'd be corralled into theme parks, at the center of movie franchises, and be central characters in thousands of children's storybooks. In the floral world, we don't need a dinosaur movie; we live in one.

The Earth's first flowers left traces of pollen in the fossil record about 130 million years ago, in the early Cretaceous. Studies of DNA suggest that flowers evolved before this, sometime between 150 million and 250 million years ago. This is astonishingly late in the evolution of life. Most of the major groups of complex animals evolved 500 or more million years ago. The first plants, none of which had flowers, came onto land at about that time, too. Flowers were latecomers, but when they finally showed up, the party really got going. Within a few million years of flowers' appearance, perhaps as few as three million, the family tree of flowering plants splayed into almost all of the main branches that are still with us today. After this initial frenzy of diversification, evolution calmed down, producing new floral diversity at a more leisurely pace.

Ancestors of water lilies and of *Amborella*, a shrub known only from the island of New Caledonia in the southwestern Pacific Ocean, were among the earliest offshoots of the family tree. Magnolias were another early branch. The palm and lily branches of the family tree came later, then came plants that would later become roses, sunflowers, and their kin. By the time fossil pollen shows up in rocks, evidence from genetic genealogies shows that all the major branches of the flowering plant family tree were established. By one hundred million years ago, flowering plants were flourishing all over the globe and, by ninety million, were dominating not only the understory where they first evolved, but forest canopies.

Truly, this was a revolution. In a few million years, the green-brown world of tree ferns, conifers, and other ancient plants was overthrown by the flowering newcomers. Plants that had dominated the land's surface for three hundred million years were suddenly in the shade of upstarts. Nothing was ever again the same. Most ecosystems are now founded on flowering plants. Our own species would never have evolved without them. We are sustained today by floral gifts, from the food on our tables to the perfumes on our skin. The Earth is a floral planet, and although we seldom realize it, we are a species almost entirely dependent on flowers.

Magnolias offer us an especially clear view of the revolutionary early days of flowering plant evolution because, although modern magnolia species all evolved in just the last few million years, they are little changed in their overall structure from more ancient times. *Archaean-*

thus, for example, a plant known from 96.5-million-year-old fossils, would not look much out of place in the magnolia grove of a modern botanic garden. Like living magnolias and their cousins the tuliptrees, *Archaeanthus* flowers sprouted from branch tips. A ring of simple petals enclosed a central axis with female and male parts. The match between ancient and modern is not perfect—the fossil flowers are rangier than living magnolias and differ in some anatomical details—but is close enough to tell us that modern magnolias and tuliptrees are, in many aspects of their biology, holdouts from ancient times. When biologists reconstruct a hypothetical ancestor of flowers by mapping floral characteristics onto a genealogy of living species, the speculative "first flower" is like a magnolia bloom, a shaggy bisexual, female atop male in a bowl of petals. For flowers, the term *bisexual* differs from its meanings among humans and refers to the presence of both female and male sex cells in the same bloom. In a bisexual flower, petals serve both pollen and eggs. This union of female and male into one structure was, as we shall see, a brilliant move, one that allowed flowers to work collaboratively with insect pollinators.

Glimpsing the earliest days of floral evolution offers more than a fun moment of time travel. We are transported to early scenes from one of the most consequential stories in evolution. When I sink my nose into a magnolia bloom, I'm face-to-face with an ancient revolutionary.

❧ ❧ ❧ ❧ ❧ ❧

Stories about revolutions often hinge on brilliant individuals and singular inventions. These "hero narratives" obscure the network of relationships and ideas from which innovations grow. Thomas Edison's light bulb was invented and perfected by a team of employees at his research lab. Printers used movable type and presses long before Gutenberg. Innovation grows from cooperation and convergence, not solitary genius. Or, perhaps, individual genius is a weaver's art, the ability to see which strands can be united and improved. The same is true for the evolution of flowering plants. No single biological invention explains how or why they changed the world. Rather, flowering plants took existing structures and ecological relationships and knit them in new patterns, stitching in a few novelties along the way.

In any complex and gorgeous tapestry, some of the needlework is showy and obvious, and other parts are raveled or knotted below the surface. For the story of the evolution of flowering plants, magnolias are the former, plainly showing the most important converging innovations. Later, we'll visit other flowers to find the secrets hidden in genetics and physiology. Magnolias, though, bring early flowering plant evolution to our unaided senses.

To better experience and understand this convergence, I pick a few southern magnolias for close study near our house in Atlanta. Although the trees are native to more southerly forests, their evergreen glossy leaves and huge flowers make them favorites here in gardens and parks. A magnolia tree planted decades ago towers over our house—this book is written in its shade—but its flowers are too high to reach. Instead, I find a scraggly row of youngsters growing between a sidewalk and a wooded gulch just down the road. In May, their branches are thick with flower buds at eye height. Each unopened bud is shaped like a candle flame, but larger, about as tall as the length of my hand. I attach fine cotton thread on the twig below the buds, then mark each thread with a unique number of knots so that I can track individual flowers.

At daybreak in late May, the air hazed with the night's humidity, I see that a flower on a low branch is waking. I call it "triple-knot" for the tags on its thread. The cupped, furred scales that enclosed the flower bud have fallen, pushed by the swelling floral tissues within. The white of the unfolding bloom stands out against the tree's dark leaves, a beacon. The flower takes the form of a champagne flute as it opens. I lower my nose to the flower's tip. Wham! Unlike the gentle wafts that greet me as

I walk past, this aroma is ecstasy. Quite literally: *Ekstasis* in Ancient Greek means to be cast out of ourselves—*ek*, outside of; *stasis*, where we stand. Even though I know that I'm just standing on an ordinary sidewalk sniffing a flower, somehow the experience knocks me sideways for a few seconds, retexturing my consciousness.

The smell is a blend of rose and citrus, with a darker undertone of cloves. Like other young flowers that I've sniffed, this one also has a tang of something green and bitter, a whiff of cucumber skin. But none of these comparisons come to mind in the moment. Instead, I'm suddenly, and only briefly, in a warm, ambrosial space. *Ek stasis*, indeed. I'm leery of admitting it—eye roll, keep a lid on the prose, flower-boy—but I'm transported.

A few moments later, I've come down. I'm a pathetic huffer lurking in my neighbors' hedging. The linguistic parts of my brain catch up, naming and judging. In the moment of wordless experience, plant chemicals lit up my sensory consciousness and emotions without the meddling influence of the fatty gray matter that clogs the rest of my cranium. Unlike the other senses, smell connects directly to brain regions that spark deeply felt experience.

Attractive floral aromas are so commonplace that we forget how strange they are in the long view of life's history. Land plants first evolved about five hundred million years ago. Animals hauled themselves out of the water about the same time and, in the many years that followed, caused dire trouble for plants. Mandibles, teeth, and stylets destroyed foliage, severed roots, drained sap, and robbed seeds. Then, flowering plants used beauty to flip the relationship. By enthralling animal senses with floral aroma and colorful petals, plants spoke directly to animal nerves. The plants turned creatures that had been mortal enemies into partners and servants. Both flowers and their animal partners benefited, surging in diversity and abundance through the Cretaceous as flowering plants rose to ecological prominence all around the world.

Flowering plants are successful and diverse partly because they are champion relationship-builders. The evolutionary trees of many of the world's most common insects—moths, butterflies, wasps, bees, flies, and beetles—fanned in twiggy profusion as flowering plants appeared and diversified. Some, like bees and butterflies, owe their existence to flowers. Bees evolved from wasps 120 million years ago, feeding on floral pollen and nectar as flowering plants took over much of the planet. The mutual reliance between flowers and bees continues to this day, and most bees feed exclusively on flowers. The need to learn which flowers are rewarding also gave bees the most sophisticated cognitive abilities of all insects, along with senses tuned to the aromas, shapes, colors, and even electrical fields of flowers. Butterflies evolved from moths about one hundred million years ago, also in response to the new abundance of flowers. Unlike moths that feed mostly at night, butterflies seek flowers during the day, often searching for them by sight rather than smell. This created new opportunities for visual signaling during courtship among butterflies, accounting for their bright colors, a contrast to the camouflage of their moth ancestors who signal to one another in the dark using scent. Other insects, like beetles, wasps, and flies, existed long before flowers, feeding on sap and pollen, but became vastly more diverse as flowers took over.

Modern birds, too, diversified in what a recent study of their evolution called "remarkable lockstep" with flowering plants. Many ancient birds fed directly on flowers and fruit, and others benefited indirectly as flowering plants built productive forests. Without flowers, there would be no nectar-drinking hummingbirds, sunbirds, and honeyeaters. More, birds that eat fruits and grass seed are wholly dependent on flowering plants, as are the legions that feed on flowering plants indirectly by feasting on insects and other herbivores. Our own dependence on grass seed (wheat, corn, and rice) and fruit (plant oils, especially) is not so different from that of the birds and, as we shall see, was a catalyst for human evolution.

The evolution of flowering plants also filled the world with animal song. As readers of my book *Sounds Wild and Broken* will know, almost all of the modern world's singers, from cicadas to crickets to grasshoppers to birds, exploded in diversity in concert with flowering plants. A planet in bloom is a planet in full voice.

Flowering plants were not the first great green cooperators. From the moment their ancestors first came onto land, five hundred million years ago, plants have been working with belowground creatures like fungi and bacteria. These relationships built the world's soil and were the foundations of the first forests, long before flowers evolved. Flowering plants took this cooperative spirit to the aboveground world. The air became as thick with relationships as the soil. Like the soil, the flowers' dense network of relationships with animals created new possibilities. Take away soil, and the Earth would wither. Take away the flowers' relationships with animals, and almost all modern animals and most ecosystems would disappear.

If this were a nature documentary, cue the soaring violins and a long shot from a drone of vast expanses of rainforest, then zoom to flower-besotted bees and fruit-plucking birds. Cue also Technicolor and aroma diffusers. Flowering plants added pigmented and scented blooms to an otherwise brown, green, and gray world that smelled of leaf, mushroom, and mud. In the words of the twentieth-century science writer Loren Eiseley, the "old, stiff, sky-reaching wooden world had changed into something that glowed here and there with strange colors." Fruits, too, dazzled and smelled delicious, and offered food in packets dense with nutrients and energy. These fruits even gave our own ancestors— ancient primates—the ability to see red, the better to find food. Today, perfumes, air fresheners, electronic screens, dyed clothes, and specially bred flowers stimulate us with diverse experiences of aroma and color. It is hard to imagine what startling innovations floral aromas and hues would have been in a previously quiet, ferny-green world. These sensory enticements knit animals to flowers, to the benefit of both.

Cross-species experiences of beauty, then, turned hundreds of millions of years of plant-insect war into fruitful cooperation. Partly so. Plants still wrestle with hordes of murderous, thieving animals, but the evolutionary genius of flowers was to convert some of these pirates into solicitous, hardworking allies. This then spurred diversification across the tree of life.

Flowering plants did not invent the experience of beauty. Animals had been sniffing one another with admiration or fear for hundreds of millions of years. Animals also used gastronomical aesthetics to seek and judge good food. The color receptors in animal eyes, too, first evolved to ogle potential mates and snacks. Flowers took these self-centered animal experiences and turned them outward, stitching distantly related species—plants, insects, vertebrates—into cooperative relationships, mediated by sensuality.

The human love of floral aroma and pigment is a curious form of voyeurism. We get a surge of pleasure from eavesdropping on plants' chemical and visual whispers to insects. We're so enamored of the experience that lately we've domesticated floral aromas in perfumes and the appearance of flowers with plant breeding. In later chapters I visit perfume gardens and flower shows to explore the beauty and hidden brokenness of these entanglements. Magnolias teach us that these aesthetic experiences of the moment carry messages from deep time. Human sensuality—sexy perfumes, lovely gardens, gorgeous bouquets—is most thrilling when the boundary between human and nonhuman blurs. This unifying power is one the flowering plants' great creations. When we spritz rosy eau de toilette or give a bunch of flowers, we're evoking the flowers' ancient spell, pleasure that reaches across species boundaries and unites unlikely partners.

❦ ❦ ❦ ❦ ❦ ❦

Over the morning, triple-knot's petals ease open, turning the flower's shape from a champagne flute into a wineglass. A column stands at the center of the flower, cupped by the petals. The column's base is sheathed by about a hundred upward-pointing white waxy struts, each the size of a matchstick. These will mature into pollen-shedding stamens, the male parts. Each pollen grain is a tiny capsule enclosing what will become sperm, the plants' way of transcending rootedness and winging its genes far and wide. Atop the tidy struts stands a conical knob bristling with about fifty downcurved yellow-green claws poking from a felt of sandy-colored hair. Each claw is a female stigma whose sticky surface traps pollen. Buried below the claws are eggs enclosed in ovaries.

Like many other modern flowers—roses, lilies, daffodils, and daisies—magnolias are biologically bisexual, making pollen and eggs in the same flower. Most ancient nonflowering plants, though, segregated the sexes. A few lived as male and female individuals, as ginkgo trees mostly still do today. Many had separate male and female parts on the same plant, as modern pine trees do with male and female cones. Sexual segregation is great if pollen is carried by wind. Many pine trees, for example, put female cones at the top of the tree to better catch pollen from far away. But the first flowering plants worked with insects, not the wind. Early flowers needed to both give and receive pollen, and to reward insects with food. A bisexual flower does all three. Bisexuality therefore enabled cooperation with insects, letting beauty do its work. Today, with some exceptions, most flowers stick with the insect-friendly bisexual innovation of their ancestors.

A dozen black-and-gray beetles crawl over the sexual parts of the triple-knot magnolia flower. The beetles are tiny—the size of rice grains—and slid through the openings between the petals as soon as the flute opened. A few beetles rummage over the tidy verticals, but most go higher, to the claws. These are tumbling flower beetles, named for their careening escape tactics. When I nudge the flower, three of them kick off with their hind legs, then spin around pointed rear ends as they fall, bouncing off leaves before hitting the ground.

Tumbling flower beetles love to eat pollen and nectar, and they are common visitors to open-cupped flowers. They are also the oldest known flower pollinators on the planet. Ninety-nine million years ago, a tumbling flower beetle made the mistake of getting caught in tree resin. Ooze engulfed the beetle. Entombed in what later hardened to amber, the beetle's body is so perfectly preserved that, today, the animal seems to have tumbled into the resin only hours ago. Flower pollen is caught in the fine hairs on the time traveler's body. Its mouthparts are clearly visible and, like its modern kin, are adapted to eat pollen.

Just like their long-dead ancestor, the beetles in triple-knot are dusted with pollen. As the beetles clamber, the claws' sticky surfaces catch some of this pollen dust. Plant DNA has taken wing. The flower's first task is done. Thanks to the beetles, pollen grains will hatch on the claws. Invisible to human eyes, pollen cells emerge and burrow through the claws' tissues toward eggs hidden in ovaries deep in the flower's core. Because the male part of this flower has yet to mature, the female part of the flower is gathering pollen from other trees, minimizing the risk of self-fertilization.

At this stage, when the female parts of the flower are collecting pollen, the flower may be duping the beetles. The beetles seek a meal of fresh pollen, yet the flower has none. Instead, the flower mimics the smell and appearance of a pollen-shedding flower. The claws look somewhat like the pollen-making stamens lower in the flower and are toughened to withstand chewing by beetles. The deception is slight—pollen is on its way—yet magnolia gives us a hint of the deceits that would come later in flower evolution, especially among the master-illusionist orchids.

I look up from triple-knot's flower and gaze into millions of leaves and twigs in the magnolia and other trees nearby. Amid this vastness, a pinch of magnolia pollen finds a stigma. Unlike Cupid's blindly shot and inaccurate arrows, pollen from another tree has, from a great distance, sunk its love dart into a target the size of a cherry, namely triple-knot's stigmas. Or, from triple-knot's perspective, the female part of the flower has reached through dense forest and grasped her hidden mate, like the goddess Eos plucking her lover from a teeming throng of mortals. The union was timely. Magnolia flowers receive pollen for only a few hours on a single day. The hunger and keen senses of beetles make pollination both accurate and efficient. The breeding season for magnolias lasts several weeks and so, although a single bloom is receptive for a short time, the abundance of surrounding flowering trees and the flower-to-flower flights of beetles mean that pollen is likely to arrive at the right time.

Thanks to a beetle trapped in amber, we know that insects have winged pollen among flowers for at least ninety-nine million years. But the flower-insect partnership is older. Most of the surviving species in lineages of early flowering plants are partly or entirely insect pollinated, suggesting that their ancestors were, too. Likewise, the magnolia-like hypothetical primordial flower that biologists reconstructed from genealogies of living species is an insect-pollinated cup. Fossilized flowers provide another line of evidence. At one site in Kansas, hundreds of flower parts are encased in 103-million-year-old stone. Many of these remains are notched, nibbled, or punctured, signs of flower-feeding insects like shield bugs, thrips, beetles, flies, sawflies, and moths. In modern ecosystems, these insects not only munch on flowers, they sup nectar and eat pollen, then wing pollen from one flower to another. Pollen from these fossilized flowers is mostly rough-surfaced and stuck into clumps, making it easy for insects to carry, unlike smooth-surfaced wind-borne pollen. Flowers, then, likely first evolved to attract and reward insect visitors.

Flowers' self-interested hospitality continues. About 90 percent of to-day's flowers are pollinated by insects. Flowering plants, though, did not invent insect pollination. Instead, they vastly expanded on what had been more happenstance relationships between plants and insects. Club moss spores are lodged in the fossilized remains of insect guts from 309 million years ago. Fossil earwig-like insects from 280 million years ago have pollen stuck to their bodies. These insects were preying on plant spores and pollen, helping themselves to portable snacks. As they fed and moved, no doubt insects ferried pollen and spores, albeit in a hap-hazard way.

Nutritious pollen was not the only reward for ancient insects. Non-flowering plants like early conifers and their relatives oozed drops of sweet fluid near the plants' eggs. These drops, still found in some living nonflowering plants like cypress and yew, capture pollen, then provide sugar for the sperm cells that hatch from the pollen grains. These polli-nation drops became energy drinks for insects. By 165 million years ago, some scorpionflies, true flies, and lacewings had evolved fluid-slurping tubular mouthparts. The plants involved, though, lacked petals, attrac-tive aromas, or other sensory enticements. They were also unisexual, segregating pollen-producing and egg-bearing parts. Pollen-feeding in-sects had no reason to visit female flowers, and sugar-drop drinkers no cause to stop on male flowers for pollen. Flowering plants used sensory lures and bisexuality to improve on these stumbling early relationships. Beauty enticed pollinators, nectar and pollen fed the insects, and bisex-uality made sure that visitors were useful to the flowers.

❦ ❦ ❦ ❦ ❦ ❦

By late morning at the triple-knot flower, the claws have dried and yellowed, their short few hours of pollen gathering over. As female receptivity wanes, the male part of the bisexual flower matures and sheds pollen. The matchstick sheath at the base of the central column loosens into a brush of dozens of upward-pointing struts. Each one is ivory with a flush of purple where it attaches at its base. The struts are now hazed with pollen, especially along their edges. Beetles crawl into the thicket. As the insects move, struts fall from the central column and accumulate in a jumble inside the flower's bowl. More beetles roam through this debris. Their mouthparts are too small to see, even with a hand lens, but the insects' focus suggests that they are eating the pollen, getting their protein-rich reward. Magnolias supplement this with nectar at the base of the flower, although these sugary secretions are miserly compared to many other species of flowering plant. Ample pollen is reward enough.

I sniff the flower again. The aroma is stronger and sweeter, with intense spicy overtones. The bitter green edge is gone. As pollen is shed, the flower cranks up its chemical and visual signaling, intensifying the aroma and throwing open the petals into a disk that is shockingly white in the daylight. Magnolias signal their maleness with copious floral perfume and a blazing-white ruff. What a contrast to our impoverished cultural imagination of sex, gender, and flowers. In contemporary Western culture, floral perfume is almost exclusively marketed as feminine. Men either shun scent or are guided to aromas of wood, moss, and dark spices. Flowers subvert such narrow ideas. Magnolia's scent—and that of roses, violets, and other perfumery standbys—is the aromatic signature of a biologically bisexual flower leaning into their male side as they shed pollen.

Compared with those many other species, magnolia blooms are especially male. We can tell this from their showiness and size. Across all flowers, those whose primary aim is sending as much pollen as possible into the world tend to be large and invest proportionally more in conspicuous, large petals. These are species in which competition among the male parts of different flowers is intense, both for the attention of pollinators and, when pollen arrives at the female part of the flower, for access to eggs. In other species, it is the success of eggs that is the primary limit on the flowers' success, not the amount of pollen that can be sent out. In these, such as the gentians and verbenas, flowers have relatively smaller petals and invest more in the protective structures around the eggs. Magnolia flowers are like elk antlers, giant displays that evolved through jousting among males.

Alongside my sniffing, I've been zapping the flower with a portable laser thermometer, checking for sexual fever. I aim the machine with the laser guide, probing the inner petals, then—*ping!*—the digital readout tells me the temperature. The process is like scanning a barcode at the supermarket checkout and feels at odds with the luscious organic union of human senses with floral aroma and color. *Ping ping ping*: The temperature is a little warmer in the floral cup, but no more so than in a protected space between leaves. In some other species of magnolia, though, the fertile flower warms above the ambient temperature by up to 10 degrees Celsius (18 Fahrenheit), especially when pollen is shed. Like perfume on warm skin, this floral heat speeds the release of aroma to the air. A warm flower shouts louder to insects. Heat can be its own reward, too. The activities of insects are often constrained by temperature. On a cool morning, they are too sluggish to fly. By resting in a heated flower, insects can break out of their cold torpor and start their day earlier. Warmth also benefits developing tissues in the flower, speeding their maturation.

It is a mystery why some magnolia species warm their flowers but others do not. Heated flowers grow on magnolia species that live in the cool mountains of China as well as species from the tropical forests of Brazil, so this feature is not restricted to just one climate zone. Perhaps the heated species face especially stiff competition from other types of plants for pollinators, making it worth their while to burn scarce energy stores? In a crowded market, you need to shout louder. The huge flowers and intense aroma of southern magnolias may attract sufficient numbers of pollinators without adding warmth.

Many other "basal" flowering plants, species that branched off the family tree early in evolution, also warm their flowers. Heat might have been an early inducement and reward for pollinating insects. Water lilies and soursop, for example, crank up their internal chemistry to warm flowers, as do a few nonflowering cycads and conifers. Today, the record holders for floral heating are more recently evolved plants, the *Philodendron* genus common in tropical forests and often grown as houseplants. In some species, flowers routinely get 10°C (18°F) above ambient, with a few exceptional examples of temperature boosts of 34°C (61°F). They achieve these floral hot flashes by peppering their flowering stalk with sterile male flowers whose sole task is to pump their metabolisms at rates that rival those of hummingbirds. These pollenless male flowers are bed warmers, not lovers. Floral heating in this species coincides with the release of aromas that attract pollinators such as scarab beetles, some of which feed and mate exclusively on *Philodendron*.

Triple-knot's white glow and strong scent last through the afternoon into the evening. As dusk falls, the blooms catch the scant light and stand out in the gloom, like paper luminaries. Hey, beetles: Here! When the beetles arrive, the flower encourages them to stay. In the center of the disk of petals, two smaller petals stood erect all day, cupping the central column. Now another rises. These inner petals form a rotunda with the reproductive column inside. In tropical magnolia species, this nighttime closure traps large beetles, thoroughly dusting them with pollen all night. But for the southern magnolia, there is no trap, especially not for the tiny tumbling flower beetles, just hospitality. The nightly closure is more vigorous in the southern magnolia's original home, more southerly coastal forests, but even in the relative cool of Atlanta nights, the blooms clench as light fades. Insects, the flowers seem to say, stay tonight in this scented bower, why don't you, safe from predators and rain? Later at night, I visit triple-knot with a flashlight. There they are, beetles in their bed of petals, some seemingly snoozing, others rummaging through the mess of pollen.

Petals are so ubiquitous among flowering plants that we forget what an innovation they were. By reshaping the ordinary green leaf into petals of various kinds, flowering plants produced a structure that allows them to enclose, support, and display their sexual parts, to scent the air with aromatic oils from petal surfaces, and sometimes to trap pollinators. Most modern flowers have two types of petal, showy inner petals encircled by green protective ones called sepals. Magnolias have just one kind, known among botanists as tepals, a simple design that is perhaps a holdout from the earliest days of plant evolution.

Petals are not just passive billboards and aroma diffusers. In some more recently evolved plants like evening primrose and petunia, petals serve as ears and electrical sensors, detecting the arrival of pollinators so the plant can sweeten its nectar or pass the message on to neighbors, perhaps to coordinate the flowers' responses to pollinators. Whether magnolia flowers communicate among themselves is, so far, unknown.

The sensory enticements of a fertile flower are ephemeral. No matter how old the plant, the flowers that it presents to the world are fresh-faced, each petal and other flower parts no more than a few days old, an alluring signal to pollinators. The flowers on a one-hundred-year-old magnolia tree are as youthful as those on a sapling. Would that our human faces could look so eternally young. Flowers achieve this bright, lively sexual display by newly growing each flower from embryonic cells in a bud. Magnolia blooms last three or four days, about an average longevity for flowers worldwide. Some asters offer flowers that last only two hours. A few wild orchids can persist for a month. In general, the more reliable your pollinators, the shorter your bloom. But floral perfection never lasts long compared with other parts of the plant like leaves and roots. Once pollen has been received and shed, flowers have no interest in charming pollinators. The flower digests itself, returning precious nutrients to the rest of the plant to be recycled into new blooms. I return to triple-knot in the morning and the ambrosial scent and gorgeous white bloom are gone. Rain has knocked the pollen-bearing struts out of the rotunda into pools of dull orange water cupped in browning petals. I sniff and the aroma is tannic, with a faint whiff of the compost heap. Despite the rain, the tips of the claws are entirely dried out and blackened as if singed. The beetles have fled.

Over the next four days, decay proceeds, hastened by chemical signals within the flower that tell cells to die and return their assets to the parent. The struts and petals turn leather brown, then fall. The aroma sours into a day-old apple core. But the "flower" is not dead. It merely sheds the baggage of youth and moves on. The top part of the column fattens by about a tenth of its volume each day. Swelling tissue engulfs the claws, leaving only their tips at the surface. After a week, the flower looks like a miniature pineapple atop a purple stalk. Inside, sperm from pollen meets eggs, and embryos start their development.

The flower was a lusty bisexual. Now the flower matures into fruit whose aims are maternal. Although the developing fruit looks radically different from the sexual parts of the flower, anatomical continuity underlies the transformation. There is no tissue in the fruit that was not already present in the flower. Even the maturing embryos inside the fruit, each one a new genetic individual, are unions of floral sex cells.

A fruit is a mature flower. The scientific term for flowering plants, *angiosperm*, recognizes this: from the Greek *angeion*, "vessel," and *sperma*, "seed." A flowering plant is one that encloses its seeds within fruits made by floral tissues, unlike all other plants that have naked or near-naked seeds. And so, when flowering plants evolved, they invented botanical motherhood. More precisely, they greatly advanced and elaborated earlier attempts at maternal care of seeds. Like flowers' partnerships with pollinators, these maternal innovations not only helped the plants, but opened opportunities for many other species. Bite into an avocado or a grape and we're eating fruit and thus flower. The starch in rice and wheat comes from another floral innovation, food supplies from the mother flower for the embryo, kept in special storage tissues in the seed. What is true for our diet is also true for ecosystems worldwide, from rainforests to coastal wetlands. Flowering plants mother and feed the world.

What does a fruit do? It encloses the growing seed. Some are wall-like, defending the seed in a husk. Others swaddle the seed with tissues so poisonous that no animal dares bite. These same tissues later turn from bane to delicacy, enlisting animals as seed dispersers. A few fruits buoy seeds across salty oceans. Some mature into catapults or guns. Many dry into wings and carry seeds aloft. All fruits nurture the seed by easing extremes of moisture or temperature, and by deterring pathogens like bacteria and fungi.

Before the evolution of fruit, seeds received less solicitous care. Pine cones provide temporary protection, but then release naked seeds to the wind or to animals: You're on your own, kiddo. Cycad seeds often have colorful and fleshy seed coats to attract dispersing animals, but this simple coat did not then evolve into the wide range of forms that flowering plants have invented. Ginkgoes independently evolved a fleshy covering for their seeds, but they founded no dynasty, perhaps because the fruit-like coat was dissociated from the flowering plants' other innovations like cooperation with pollinators.

Beauty, bisexuality, and motherhood: The magnolia reveals a powerful convergence. Each had precedents among earlier plants. In a brilliant move, the first flowering plants united these innovations into a single structure, the flower.

I check in on triple-knot throughout the summer. After the hour-by-hour excitement of flowering, things calm down. The fruit grows week by week at a slower pace than its initial swelling. After a month, the blackened tips of the claws poke from bumps of tissue. Each bump looks like the pad on a cat's paw and is made from the flower's ovary tissues. Now these tissues enclose one or two maturing seeds.

In mid-August, bumps flush pink and a vertical furrow appears down the center of each. In the second week of September, the furrows split and a bean-sized seed erupts. The seeds' skin is smooth and bright crimson, a shock to the eyes. A few seeds lodge in the splits. Most fall out and dangle by a silvery thread, swaying and flashing red in the wind. By the end of the month, the fruit has browned to a dry husk, a muted ground against which glow the seeds.

Red is a popular autumnal fruit color here in the southeastern United States, proffered by magnolia but also dogwood trees, spicebush shrubs, and wildflowers like jack-in-the-pulpit. The plants hope to attract birds, another ancient trick of flowering plants. Bird eyes, like those of their reptilian kin and unlike most mammals, have a light receptor tuned to orange and red. The color speaks of sexiness and social status in many bird species. By daubing their fruits or seeds in red pigment, plants tapped into this preexisting visual enthusiasm. A few nonflowering plants like cycads advertise their seeds with orange and red, but flowering plants were more exuberant, like kids at a painting party. They splashed red onto seeds, fruits, and petals, and offered ample sugary and fruity treats.

Today, a large magnolia tree with mature fruit in Atlanta can host four species of woodpecker and three species of thrush, all at the same time. Each tree offers hundreds of fruits and thousands of seeds, and so there is plenty to go around. Triple-knot is only about a decade old and, although the trees grow fast and live only about a century, it will take another thirty or more years before this tree reaches the grandeur of a flower-festooned giant. For now, a single downy woodpecker probes its fruits. This is an old relationship. When magnolias first evolved, birds had been around for at least fifty million years. We don't know the color of the ancestral magnolia seed, but a good guess would be crimson red.

❦ ❦ ❦ ❦ ❦ ❦

That the early days of flowering plants should be reenacted in an everyday setting is wondrous. But magnolia tells a peculiar story, one at odds with the narrative of almost all other flowering plants. Magnolias are a rare time capsule of animal relationships and floral forms mostly unchanged. For other flowering plants, early innovations led instead to explosive diversification and rapid change. This profusion of flowers catalyzed and fueled evolution in animals, microbes, fungi, and other plants.

Magnolias are holdouts, radicals in their youth, but now traditionalists coasting on an old formula for success. Like former firebrands now turned conservative, they have some great tales to recount about the past. Most other flowering plants, though, did not settle down after an exciting youth. Instead, they continued the revolutionary spirit, remaking the world with their creativity and productivity. For them, what happened one hundred million or more years ago was prologue. To understand these restless others, we need to turn our gaze from the magnificence of the southern magnolia trees to the weeds poking their impetuous heads from field margins and pavement edges. They have a secret about ancient innovation buried in their DNA and a message about who will thrive on a future Earth.

Goatsbeard

Abominable mysteries and

genetic storms

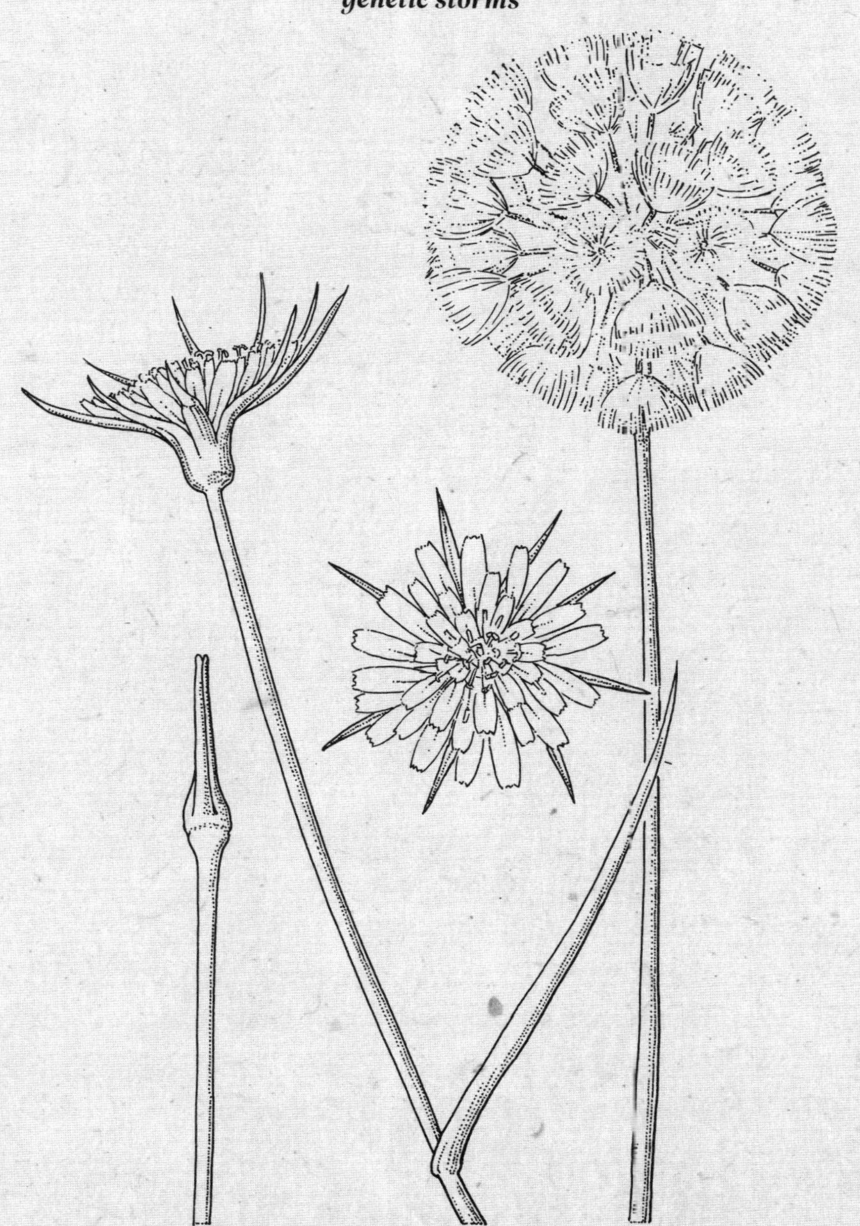

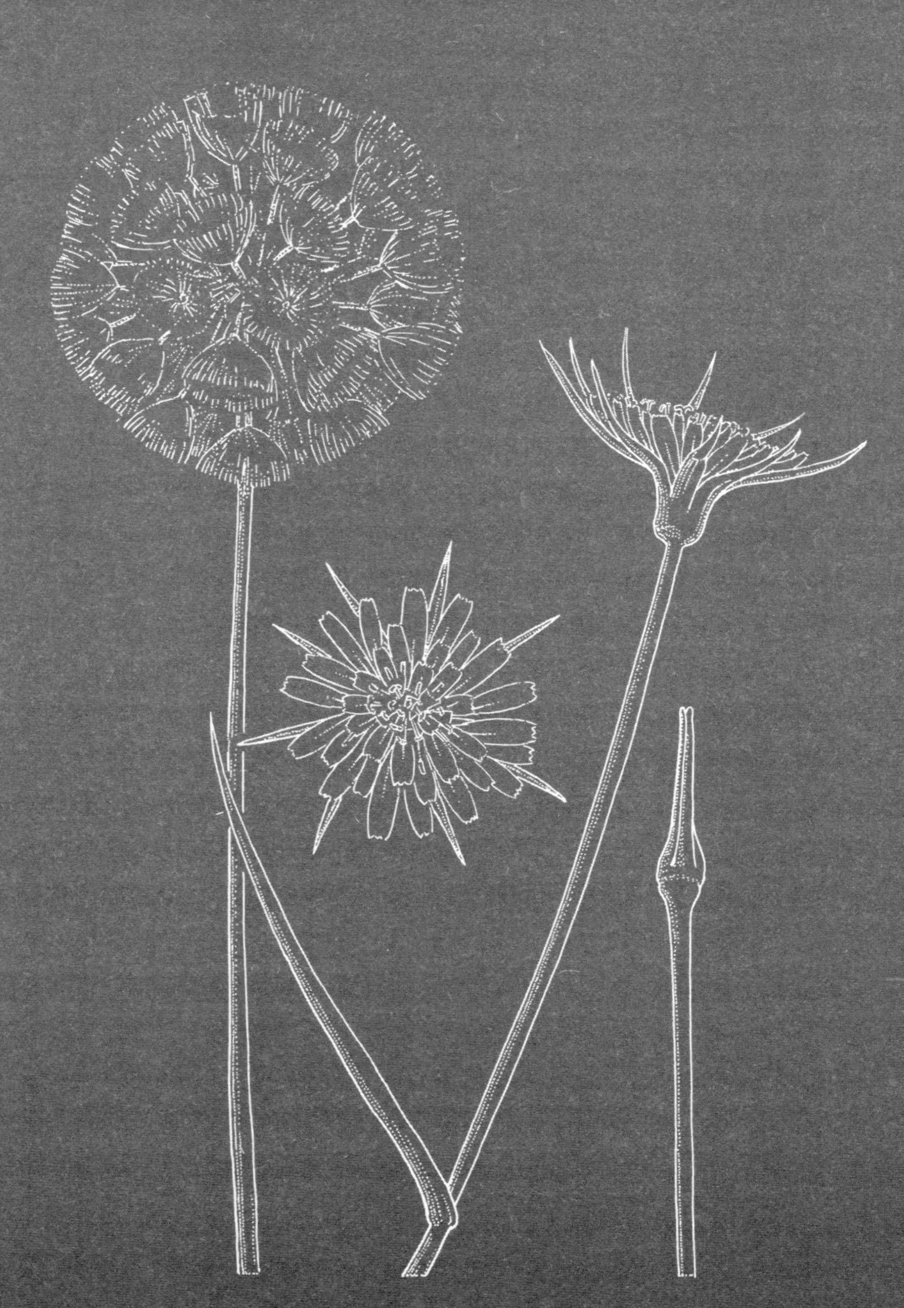

YELLOW GOATSBEARD IS A BADASS DANDELION. More firework than pom-pom. The sulfur-yellow bloom sits at the center of a burst of green spikes. The yellow outer petals are long, too, and tipped with tiny teeth. The flower stalk reaches calf high. The overall effect is of a ball of sunlight lancing up then exploding. The flower matures into a fluffy seed head sometimes as big as a tennis ball. It takes a vigorous huff to dislodge the seeds, each of which floats from a tufted, silky parachute as big across as my thumbnail.

I first met goatsbeard as a "weed" sprouting from the untended edges of the community garden in Colorado where my partner, Katie, and I grew vegetables and flowers for a few years. Katie is a violinist and has grown flowers and vegetables for years, with a musician's keen appreciation of the beautiful tension in gardens between order and surprise. Goatsbeard fits the latter, poking up on the margins of our garden plots and blooming along the edges of roads and well-worn hiking trails. The species is not native to the region and is often regarded as an undesirable interloper, but we both admired how the flowers erupted from gravelly, beaten-up places that other plants shunned. Especially during the first months of the COVID-19 pandemic, when Colorado's skies were often shrouded in smoke from massive forest fires, goatsbeard offered unexpected exclamations of beauty in a somber world.

Originally from Europe and western parts of Asia, yellow goatsbeard is named for its spiky flower and hirsute seed head, each somewhat caprine. Other plants go by the same name—notably the shrubby goatsbeards in the rose family—but these weedy, dandelion-like goatsbeards also invoke animals in their scientific name, *Tragopogon*—billy goat, *tragos*, and beard, *pogon*. In its homelands, the taproot is sometimes eaten as a vegetable and the leaves used as a potherb and a medicine for inflammation. Goatsbeard also goes by the name oyster plant, for the taste of the cooked root, and salsify, a name originally from Italy.

As Europeans colonized other lands, goatsbeard went with them. The plant now grows on all continents except Antarctica. Goatsbeard has done especially well in the American West, where it thrives in the disturbed edges of roads, railway lines, and cultivated fields. Wherever humans churn the soil, goatsbeard arrives on its parachute and sinks its taproot. Unlike some other non-native species like kudzu and ivy, goatsbeard seldom crowds out other species.

Immigration can produce strange and fruitful convergences. Two species from the same genus as yellow goatsbeard also traveled to North America from Europe, the purple and meadow goatsbeards. The purple species is the one most favored by cooks for its root. Back in their original homes, the three goatsbeards occasionally grew alongside one another, but mostly lived separated by their preferences for climate and soil. In western North America they bloomed together, the three species side by side in the same railroad beds and roadsides. Like the Jets and the Sharks in *West Side Story*, this juxtaposition of immigrant communities led to a remarkable love story. The goatsbeard romance reveals a genetic secret about evolution, one that tells us how the first flowering plants on Earth became so successful.

* * * * * *

"I have been having an exciting time this spring studying the origins of species in *Tragopogon* through natural hybridization," wrote Marion Ownbey to a colleague in 1949. Ownbey was a professor and director of the herbarium at Washington State University. Goatsbeard caught his eye after he moved in 1939 to the Palouse region of eastern Washington state.

When Ownbey arrived, the Palouse had been under the plow for less than a century. Following the armed takeover of the region from the Coeur d'Alene, Palouse, Spokane, Cayuse, and Nez Perce peoples by white settlers in the 1850s, the deep, rich soils of the former prairies were turned to growing wheat, a practice that continues to the present day. The rolling hills of the region are plowed edge to edge and planted in huge fields of wheat, along with some chickpea, one of the world's most productive agricultural granaries. Goatsbeard arrived with settlers' vegetable gardens and as a weed in sacks of imported grain.

In a scientific paper in 1950, Ownbey explained that yellow goatsbeard had "successfully invaded waste places, roadsides, fields, and pastures" and was common throughout the region, having arrived in the early 1900s. Purple and meadow goatsbeard, though, were "almost wholly confined to towns," presumably blown onto roadsides from gardens gone to seed. But among these familiar species were newcomers, giants that did not seem to fit within any known species. Intrigued, Ownbey searched the region for goatsbeard, seeking patterns and explanations. As he explored, he collected goatsbeard plants, laid them between sheets of paper, then pressed and dried them. Stored in dry cabinets, pressed plants can keep for centuries. Like frames in a stop-motion film, each preserved plant was a snapshot of evolution. What Ownbey found was the origin of new species.

Seventy-five years after Ownbey started his research, I visit the herbarium at Washington State University to see these frame-frozen plants. As I prepare for my visit, I feel like a pilgrim, about to visit a secular, scientific shrine. For biologists, the specimens that Ownbey carefully preserved in the university's herbarium are like religious relics, carrying meanings beyond their insubstantial material presence. These objects invoke not unearthly deities or saints, but the generative powers of genetics and evolution.

Today, the herbarium sits on the fourth floor of a mercifully well-air-conditioned science and engineering library building. I visit in July when the air outside is oven-hot and dry, and tastes of smoke from forest fires to the south. Walter Fertig, the manager of the herbarium, welcomes me to a large open-plan space, worktables arranged next to hundreds of tall metal cabinets. Students bend over a few of the tables, quietly focused on plants and computers. Each cabinet holds hundreds of dried plants and lichens, four hundred thousand individual specimens in all, a record of field collections in the region from the early 1900s to the present day. The space feels cool, quiet, and orderly, a contrast to the lush profusion of plant communities outside.

"They're through here. I've pulled the *Tragopogon* types." Walter leads me to a back room, lined with bookshelves holding botany manuals. Two large cabinets stand at the rear. Their doors are labeled "WSU Types," a repository for the "type specimens" that scientists use to describe new species. Paper folders are stacked on the cabinets' shelves, each folder containing one plant or lichen, the ur-specimen for a species. In biology, materiality matters. Naming a new species involves not only a description in a technical journal, but also the killing of an individual of the new species to be put into a permanent archive. Like some religions, science also involves ritual sacrifice. The preserved individual is known as a "type specimen." Future scientists refer back to the "type" as they discover new living populations of unknown identity or develop analysis techniques undreamed of by their forebears. Evolution does not, though, revolve around fixed types. Every individual differs from others in its population, and species are always in flux. The type specimen persists in biological research because it is useful, a stationary point that reduces confusion, but this utility does not reflect the varied, changeable, and sometimes confounding nature of real life.

"Here's Ownbey's *miscellus* and, over here on the left, is *mirus.*" Using scientific names for two goatsbeard species, Walter points to two large paper folders on a wide worktable. He opens them, revealing dried, pressed goatsbeards inside. The paper and plants smell of old dry hay, warm and slightly sawdusty. The plants are flattened and their stems broken to fit onto the sheets. Purple pigment still glows from the petals of the squashed bloom of one of them. These are the plants that Ownbey collected when he described two new species of goatsbeard. Each sheet bears a label, in the distinctive print of a manual typewriter: "fertile bottom land, Pullman. June 9, 1949" and "fertile bottom land along rail-road at west edge of Moscow. June 9, 10, 1949." One plant came from Pullman, Washington, and the other from Moscow, Idaho, places just a dozen kilometers apart.

"I've never found the leaf characteristics that helpful. Or perhaps I'm just not experienced enough with these. *Miscellus* is meant to have curled leaf ends, but I often don't see it in the field." Walter speaks like taxonomists everywhere, combining skepticism about whether published descriptions capture the identities of wild creatures with self-doubt about his abilities to see the patterns that others have described. Most knowledge is tentative, especially for the identities of plant species. "The flowers are much more distinctive. *Mirus* has a bull's-eye pattern and *miscellus* is all yellow and quite large. On the hill down from the Walmart parking lot I see them in bloom, side by side. Really clear, as long as they're in bloom. After that, it gets hard." With plants, most knowledge is local, too.

When Ownbey collected and described these plants, the species they belonged to were not only new to science, they were entirely new to the world. His herbarium records captured evolution in action. The three imported goatsbeard species—yellow, purple, and meadow—had interbred, spawning giants, some of which were new species. These offspring could only breed among themselves. They were forever separated from their parent species. Instant evolution.

A biological "hybrid" is the result of a cross between two species that otherwise seem distinct. Garden strawberries, for example, are the hybrid offspring of North American and Chilean species of strawberry, an unlikely union achieved during an eighteenth-century breeding experiment in France. No horticulturalist messed with goatsbeard in Washington state, though. Instead, bees stirred the genetic pot. With three immigrant goatsbeard species growing together, pollen inevitably jumped across species lines. Ownbey found some lovely hybrid goatsbeards, including a few that made bicolored flowers, drawing from the pigment palette of both purple and yellow parents. These hybrids were tall and flowered profusely. But their flowers mostly came to nothing, withering instead of maturing into puffy seed heads. Sterility is a common fate for hybrids. A mule might be strong, but it cannot successfully breed.

In four small colonies of goatsbeard, though, hybrids were not just large, but gigantic. They had thick, succulent leaves and stems, and big flowers. The pollen grains were twice the volume of those of the parents. Astonishingly, these titans were fertile. Their flowers matured into spectacular fruiting globes. The giant hybrids could breed. The plants that Walter spread on the table for me to study and admire were some of the fertile giants collected by Ownbey from these colonies.

To find out why some hybrids were sterile but others fertile, Ownbey plucked anthers, the parts of the flower that make pollen. He squeezed anther cells between two sheets of glass, then added red dye. Peering through a microscope, chromosomes came into view.

A chromosome is a long strand of DNA spooled around proteins. All plants, animals, and fungi organize most of their DNA this way, like knitters keeping yarn in tidy skeins. Sex complicates this arrangement: Which parent gives which skein to the new embryo? The answer for many species, humans and goatsbeards included, is for adults to carry two copies of every chromosome. Parents give half of their collection to every sex cell. When egg and sperm unite, the full complement is reassembled. Humans usually have twenty-three pairs of chromosomes. Parental goatsbeards have six pairs.

The giant goatsbeard hybrids, though, have six quadruples. A mistake in cellular division doubled their chromosomes, giving them four copies of each. This doubling seems freakish, but we now know it is quite common in evolution, especially among plants. Duplications of chromosomes also often presage major evolutionary transformations, including the origin of flowering plants.

For day-to-day running of a plant or any other creature, four copies of every chromosome is excessive. Who needs four copies of a blueprint? Two is plenty to run the cells and to make eggs and pollen. Carrying excess chromosomes is costly. The DNA and protein in chromosomes divert energy and material from other projects and the excess material bulks the cell. This is likely why the fertile hybrids are so big. Their cells must swell up to accommodate an outrageously large collection of chromosomes. This excess has one great advantage: It allows the hybrids to make viable eggs and pollen.

In the sterile hybrids, those with just two copies of every chromosome, the delicate chromosomal dance needed to make eggs and pollen is a disaster. Normally, pairs of chromosomes lie next to one another during these cellular divisions. But in the sterile hybrids, the snuggling goes awry. Each chromosome in the pair comes from a different parent species and has a slightly different shape and chemical properties. The mismatch usually results in dud eggs and pollen. By doubling their chromosomes, the fertile hybrids dodge this problem. Their four copies comprise two from each parent species. When it comes time for the chromosomes to align, each parental type finds its own kind and there are no mismatches.

Doubling the chromosome count had one more immediate consequence. The hybrid offspring could no longer breed with the parent species. The different chromosome counts made union of hybrid and parental eggs and pollen impossible. The hybrids were able to breed with hybrids of their own kinds, but with no others. Two hybrid crosses successfully produced new species, yellow x purple and yellow x meadow. The purple x meadow cross yields only infertile offspring. Why yellow goatsbeard can successfully cross in this way is not known. Even among yellow goatsbeard plants, only a few can pull it off, presumably because they possess a genetic quirk that tolerates or encourages hybridization and chromosomal multiplication.

Ownbey welcomed the new species by naming them: *Tragopogon mirus* and *Tragopogon miscellus*, from *mirus*, "amazing," and *miscellus*, "mixed." *Mirus* has two-toned flowers, a yellow eye in a ring of purple. *Miscellus* has big yellow flowers, with long petals. In the decades since Ownbey's discovery, these amazing mixed goatsbeards have thrived. His searches yielded four "small and precarious" populations, all made up of only a few dozen individual plants. Today, the hybrids grow across eastern Washington and parts of Idaho, sometimes in populations of thousands, and they are pushing out the parental species, especially meadow and purple goatsbeards. Botanists have also found the new species in Montana, Wyoming, and Arizona. Some of these populations descend from the first ones found by Ownbey. But most are the result of independent hybridizations, showing that the parental species continue to cross and found new fertile hybrid populations. Genetic studies show that *Tragopogon mirus* has formed at least thirteen times and *Tragopogon miscellus* perhaps as many as twenty-one times.

The hybrids are already evolving. Chunks of DNA have moved among chromosomes. Seventy percent of hybrid plants have lost or gained a chromosome. Some genes have been deleted or silenced. Goatsbeard takes two years to flower, and so only forty plant generations, at most, have passed since these species appeared. Natural selection is wasting no time.

Hanging on the side wall in the type specimen room in the WSU herbarium is a framed black-and-white photograph, barely bigger than a sheet of printer paper. Fifty white men in 1940s overcoats and ties stand in orderly rows, staring unsmiling at the lens in front of an ivy-clad brick building whose gutters are filled with snow. Someone has taken a fine black pen to the glass over the photograph and scrawled names next to some faces: "Dobzhansky, Wright, Fisher, Mayr, Simpson, Haldane." These are the men who shaped early twentieth-century biology, gathered in January 1947 in Princeton for a conference on evolution. Behind the patriarchs in the front rows stands a young Ownbey.

At the 1947 conference, talk was of geologic time and the emerging sciences of evolution and genetics, although the discovery of DNA was yet to come. The written record shows no evidence that Ownbey gave a formal presentation at the meeting, but he was present, listening to the latest ideas about evolution. How, for example, might new species appear? Darwin named his famous book *On the Origin of Species*, but a century later, biologists had few concrete examples. Might new species appear by interbreeding of closely related parent species, perhaps accompanied by a doubling of the chromosomes? Some biologists thought this likely, especially among flowering plants. But no one had convincingly seen this happen in the wild.

Ownbey had first documented sterile hybrid goatsbeards on the roadsides of Pullman in 1946. He returned from the 1947 conference with new ideas and new eyes. In the spring of 1949, he found the fertile hybrids, writing to his colleague, "This spring I found their amphidiploid [fertile hybrid] and with this lead worked out the rest of the story. It is easily the most spectacular case that has yet been discovered." He then wrote what is now his classic 1950 paper describing the rapid appearance of new goatsbeard species in the Palouse. Biologists had their example of the origin of species, in real time. The key to the whole process was hybridization and doubling of the chromosomes.

Today, we know that such doublings not only explain recent origins of individual species, but they were also the hidden inner revolutions that allowed flowering plants to first evolve and explode in diversity. The floral planet exists partly because chromosomes multiplied within mutant ancestors.

In a letter to the botanist Joseph Hooker in 1879, Charles Darwin wrote, the "rapid development as far as we can judge of all the higher plants within recent geological times is an abominable mystery." By "higher plants," he meant the group that includes most modern flowering plant species such as sunflowers, orchids, and roses. These plants suddenly appeared in the fossil record in great profusion, unlike plants Darwin and his contemporaries believed to be more primitive, like grasses. Darwin's plant taxonomy has been superseded—grasses are no longer considered to be more ancient or primitive than sunflowers—but the abominable mystery remains. Flowering plants burst onto the scene and quickly rose to prominence. An explosion with an unknown cause.

With hindsight, we can now see that Darwin could never have solved this mystery. He was right to feel distraught by what, in the same letter to Hooker, he called his "wretchedly poor conjecture" about the origins of flowering plants. Darwin lacked all knowledge of modern genetics and could not see the secrets hidden in plant DNA.

Recent studies of plant genomes—the term used to describe all of the DNA in a cell—uncovered shocking scars in the DNA of modern plants, the genetic marks of ancient overhauls, analogous to finding missing chunks or strange rearrangements in a modern book, the signature of past printing or proofing errors. When geneticists map the DNA of living flowering plants, they infer from the arrangement of the DNA sequences that, like goatsbeard hybrids, ancient flowering plants doubled all their chromosomes. This wholesale copying and expansion was followed by millions of years of pruning, a process only just started in goats-beard. The ancient plants with doubled genomes eventually cut back their DNA so vigorously that they got back to just two copies of each chromosome.

Copy All. Paste. Delete, selectively. What possible advantage could there be to going through such a process? Genome-duplicated plants have four copies of every chromosome. This means four copies of every gene. There is wiggle room here for innovation. If one of the four copies mutates, the plants still have three "normal" copies, more than enough to keep cells happy. The mutant fourth gene becomes an explorer. Perhaps the mutation causes the gene to make a new form of protein, one that confers frost resistance or evokes a novel aroma in flowers. Or the mutated gene might shift how it converses with other genes, changing the managerial network so that leaves emerge earlier or flower petals grow in a new shape. Of course, most mutations are useless or harmful. Natural selection purges these bad ideas. But natural selection also needs a source of helpful novelties. Duplicating whole genomes vastly increases the supply of genetic variation.

Unlike the plodding letter-by-letter editing that is natural selection's everyday work, the duplication of a whole genome is like a cross between a corporate brainstorming retreat and a rave. The usual rule that every gene must be productive at all times is suspended and the whole body explores new ideas and ways of being.

One result of duplication, especially when it happens several times over tens of millions of years, is that a single gene can multiply into a large family of genes. For example, most flowering plants have up to fifty versions of a gene that makes sugar-pumping proteins. Algae have four or fewer copies. Over the hundreds of millions of years that have elapsed since flowering plants last shared an ancestor with algae, repeated rounds of duplication multiplied the number of copies of these genes in flowering plants. In modern flowering plants, each version of the gene has a specialized role. Many shunt sugars out of leaf cells into the plants' veins. Others supply developing pollen or seeds. A few move sugars inside the cell. Some shuttle hormones. These genes did not each independently evolve from scratch. Instead, a single ancestral gene duplicated, and natural selection tweaked the copies into dozens of variants, plus unmodified duplicates to give especially important genes a boost. The genome plagiarizes itself through gene duplication. Composer Igor Stravinsky said of Vivaldi, "[He] is greatly overrated—a dull fellow who could compose the same form so many times over." In evolution, as in music, this can be a winning way to innovate.

As genes within a family adapt to their different roles, their DNA codes diverge. We can use this drifting apart to build a genealogy of genes within a family. Those that duplicated long ago tend to have very different DNA codes. Those that duplicated just a few years ago remain quite similar. By building such genealogies for thousands of gene families, geneticists can gaze back into time and estimate when entire genomes doubled. Darwin would have twirled his beard in delight at the results of these studies. Ancestral plants doubled their genomes 319 million and 192 million years ago, give or take a few million. That's before two of the most significant events ever in plant evolution: the origins of seed plants and flowering plants.

The "abominable mystery" is partly explained by duplication of genomes. The more ancient duplication came before plants invented the seed, a structure that is now used by all living plants except the mosses, ferns, liverworts, and kin. A seed is a nursery, larder, and protective case for embryonic plants. The later duplication, 192 million years ago, preceded the invention of flowers and ushered a burst of diversification that eventually produced most of the plant species alive today.

Can genome duplications really trigger such revolutions? The fact that duplication occurred before two major bursts of evolutionary creativity is suggestive but is hardly proof. Geneticists might have uncovered a coincidence and nothing more. The time scale, too, is pretty loose, leaving room for exaggerated conjecture. We don't know exactly when the very first flowering plants evolved. Many studies point to sometime around 192 million years ago. But the explosion of diversity among flowering plants did not happen until 130 million years ago and later. If duplication ignited an explosion of diversification, the process had a long fuse.

The evidence for the catalytic effect of duplication gets stronger if we look at what happened after flowering plants evolved. Duplications happened hundreds of times, not just at the origin of flowering plants. In most of these cases, duplications were followed by bursts of innovation and diversification, often with a lag of twenty million years or more. The sunflower family, for example, is today the second most diverse family of flowering plants after the orchids, with about twenty-five thousand living species, including goatsbeard. An ancestral duplication inaugurated the family. The same is true for the "monocots," a larger grouping that includes families such as the orchids, lilies, grasses, and palms.

Of just over one hundred ancient duplication events studied by geneticists, about two thirds are followed by increases in species diversity. These are all duplications from the distant past, millions of years ago. More recent duplications, like those in goatsbeard, were also productive. One quarter to one third of all flowering plant species may have formed by genome doubling.

We humans depend on the fruitfulness of doubled genomes. Most major modern human crop plants have doubled (or tripled, quadrupled, and more) genomes, including wheat, oats, potato, cotton, quinoa, peanut, cabbages, coffee, and sugarcane. For many, the wild progenitor of the crop plant has an undoubled genome. Although our ancestors were ignorant of chromosomes, they were skilled at working with flowering plants to craft new, especially nutritious plants from existing wild species. Often they did this by crossbreeding species, producing novel offspring that we now know had multiplied genomes. Modern bread wheat, which today accounts for 95 percent of the global wheat crop, for example, is a descendant of two rounds of hybridization, uniting the genomes of three wild grass species into one. When you bite into a bagel or a biscuit, you're eating the remains of cells that had six copies of every chromosome instead of the two copies of their ancestors. Other crop plants, like some potatoes, did not hybridize, they simply doubled their chromosomes, opening new avenues for genetic exploration. Our plant-breeding ancestors interwove their needs and desires with the wild processes of genome duplication and hybridization, creating new crops. The flexibility and innovations of plant genetics not only transformed the planet millions of years ago as flowering plants took over much of the world, these processes are the foundation of agrarian civilizations.

Goatsbeard is exceptional not because it doubled its chromosomes, but because it was caught in the act. Seldom do biologists see this process unfold before them. Most doublings lead to dead ends. But a few found dynasties and presage major innovations. Without genome duplication, many modern plant families and species would not exist. And we humans would be very hungry indeed.

❧ ❧ ❧ ❧ ❧

The goatsbeards that arrived from Europe in the American West in the late nineteenth and early twentieth centuries experienced a climatic shock more rapid and profound than any experienced by their recent ancestors. Even the end of the ice age in Europe had taken centuries. For the immigrant goatsbeards, one year they grew in a meadow in Italy or a garden in France, then the next year, America. A few seeds were deliberately plucked and shipped west to start vegetable or medicinal gardens. Others were caught up in the grain harvest from European fields, then bagged to be sent across the Atlantic. America was profoundly different from the plants' original homes. Goatsbeard that was carried from the barley fields of central Europe to those of Washington state, for example, suddenly experienced colder and snowier winters, hotter and drier summers, more sunshine, and higher winds. Seasonal rhythms were different, too, with wilder swings of temperature and moisture. The translocation also coincided with an agricultural revolution. Goatsbeard had to contend with newfangled agricultural chemicals and engine-driven plows. Then, when the plants flowered, their eggs and pollen mixed with other goatsbeard species, a sexual novelty unknown in most of their homelands. Upheaval, from all sides.

Stress makes plants mutate. Extreme heat, sudden freezes, and other environmental hardships wound the genome. Chromosomes don't divide as they should and the DNA code gets peppered with errors. These changes degrade the vitality of most plants. But in a few individuals the genetic chaos leads to useful innovations. Out of turmoil, a few good ideas emerge.

Goatsbeards' stressful new environment provoked chromosomal doubling. A few of these genome-duplicated plants thrived. Compared with their parent species, many can photosynthesize faster. The upstarts seem better adapted to the new home and are taking over. In other flowering plant species, lab and field experiments show that genome-doubled plants are often highly tolerant of drought, salty soils, excess heat, and other stresses.

When the outside world gets tough, inner transformation can help. This was true millions of years ago, too. The meteorite- and volcano-induced cataclysms at the end of the Cretaceous, for example, wiped out 70 percent of plant species. Among the species that not only survived but thrived were the ancestors of the grasses, plants that now build prairies, steppes, and agricultural ecosystems. As the world convulsed and most species died, grass ancestors doubled their genomes, made it through the rough times, and then flourished. These grassland habitats would, much later, be both the stage and spur for early human evolution.

We humans got our start thanks to the genetic resilience of grasses that made it through a world-changing calamity. Other survivors of the end-Cretaceous mass extinction, including the ancestors of legumes and bananas, also thrived after global upheaval by doubling their genomes. What's past is prologue: In our modern era of environmental crisis and transformation, the genetic flexibility of flowering plants will be just as important as it was during the Earth-changing convulsions of ancient times. The flowering "weeds" growing in pavement cracks and field margins—goatsbeard, along with others that we'll meet later—are inheriting the changed Earth.

Genome doubling provides raw material for evolutionary creativity. But doubling alone is insufficient. *Copy All. Paste. Delete*, selectively. The last step is vital. Repeated doubling without trimming bloats the genome. Chromosomes are made of precious protein and DNA, and they take time to build. Chromosomes are bulky, too. Like human hoarders, plants face a challenge after duplications. Yes, those piles of DNA code might be useful one day. But right now, they threaten to sap your energy and bury the house in junk.

The garbage throw-out starts immediately after doubling. The new hybrid species of goatsbeard are only a century old, yet by one estimate 20 percent of their genes have lost one or more copies. Add up these changes over millions of years, and doubled genomes can be drastically reduced in size. In plants that duplicated long ago, entire chunks of chromosome are deleted, all through the random processes of genetic mutation and fumbled chromosome division. Flowering plants that today have just two copies of every chromosome are all descendants of ancestors that had many more, then cut back this swarm to a more manageable collection.

Mutation may be random, but which mutants survive is not. Looking back at the auspicious doubling of the genome about 192 million years ago, we see that the genes most relevant to flowering were retained, often in multiple copies. These winning genes control the timing of reproduction, the growth of flower tissues, and the development of the fruit and seed. Most of the other doubled genetic material from that time was tossed out.

Doubling the genome, then selectively trimming, gave the flowering plants abundant new material for evolutionary exploration. There is evidence that other plants, like the ancestors of modern ferns, also doubled. But they did so infrequently and their cutting back was less thorough. Adder's-tongue fern, for example, has 1,440 chromosomes inside every cell, a clunky way of running a plant. The fork fern from New Caledonia has the largest genome of any known living cell, fifty times larger than ours, with eight copies of every chromosome, each one bloated with duplicated and untrimmed DNA. It was repeated doubling *and* zealous pruning that gave flowering plants their genetic advantage. Our animal ancestors, too, may have benefited from an ancient doubling. The genomes of all mammals, reptiles, birds, and amphibians bear the mark of two ancient duplications followed by editing, dating to the time when land vertebrates were separating from their fishy ancestors.

❦ ❦ ❦ ❦ ❦ ❦

The hard pruning of genes that followed genome doubling gave flowering plants one more advantage over others. Trimmed genomes allow cells to be small. Compact cells can feed faster.

Plants use sunshine to daisy-chain carbon dioxide into sugars. They eat air and light. There are a few exceptions to this rule, like the fungus-feeding orchids we'll meet in the next chapter. But for the vast majority of plants, air and sunlight are food. To do this work, leaf cells need water and carbon dioxide. Water arrives through leaf veins. Carbon dioxide flows through breathing pores. If leaf cells are small, the vein network can be dense and breathing pores abundant. Like chefs with the pantry and sink within arm's reach, small-celled leaves can crank out food. Plants with small cells can photosynthesize three or four times faster than their large-celled kin. But when chromosomes double, they bulk the cell. Hence the importance of trimming.

By jettisoning genetic baggage after doubling, flowering plants got to pick out a few useful doubled genes, then get back to feeding by sunbathing. This two-step process, duplicate-then-cut, was essential to the early success of flowering plants.

We can read the story in fossils. Long ago, leaves fell into mud, pressing their shapes into what became stone tombs. Cracking open these mausoleums tens or hundreds of millions of years later, we see not only the shape of the leaf—this one is like an aspen, that one a pine needle—but all the exquisite details of their veins and breathing pores. The density of veins in the fossilized leaves of flowering plants tripled over the Cretaceous. In contrast, the vein density of ferns, conifers, and other nonflowering plants was low and did not budge. Likewise, the leaf pores of flowering plants became two or more times denser than most nonflowering plants. Today, the average flowering plant has four times the vein density of nonflowering plants. Pore densities show a less clear trend, but are almost always higher in flowering plants. Keeping cells small was a great strategy for flowering plants.

Are the living nonflowering plants, the ferns, conifers, and their relatives, deficient or doomed? No, they are well adapted to homes where vein and pore density are not the main challenge, places like the understory of shady woods or the snowy boreal forests. Elsewhere, though, flowering plants feast on air and therefore abound in green profusion.

Efficient feeding is especially important when conditions get tough. One hundred million years ago, carbon dioxide levels were very high, above 1,000 parts per million, nearly four times higher than the recent pre-industrial level of 280 ppm. For plants, the air was a rich soup and photosynthesis was relatively easy. Carbon dioxide levels then halved (or more) over the next thirty-five million years, a period that coincides with when flowering plants became common. The drop in carbon dioxide was likely caused by chemical absorption into weathering rocks and deep ocean sediments. For plants in the Cretaceous, the air turned from nutritious soup to meager broth. But, unlike their ancestors and non-flowering kin, flowering plants now had more spoons in the soup bowl. Their ample breathing pores and dense networks of leaf veins kept water and carbon dioxide flowing through their leaves.

Revolutions are sometimes founded on tiny things. What could seem more obscure and irrelevant to grand narratives of Earth than pores and veins on leaves? Collectively, we even call leaves mere "litter," instead of the world-changing wonders that they are. Yet, changes in the densities of plant veins and breathing pores partly caused one of the largest over-hauls of terrestrial life, the rise of the flowering plants. Not only that, these details of leaf architecture changed the climate and created new ecosystems that are now the most important habitats on Earth. Leaves with dense vein networks and abundant pores lose a lot of water to the air. As the plants' breath rises heavenward, clouds form and downpours follow. By the end of the Cretaceous, this intensified relationship be-tween plants and sky created tropical rainforests. This was helped by the flowering plants' invention of new, open-bored plumbing vessels that sped the flow of water from roots to tree trunks. As paleobotanist C. Kevin Boyce and his colleagues put it in a scientific paper, flowering plants "put the rain in the rainforests." These wet forests are now fa-mously the most biodiverse places on the planet, and they stabilize global climate and rainfall. All this diverse living glory depends on the particular forms of filigreed plant leaves, made possible by cells made svelte by their trimmed genomes. Tiny things, indeed.

❈ ❈ ❈ ❈ ❈ ❈

Slashed-back genomes gave flowering plants leaves that could gobble light and air. The chromosome duplications that preceded this streamlining catalyzed genetic innovation. But to what end? Evolution is not a magpie, attracted by twinkling novelty for its own sake. For natural selection, the measure of creative worth is utility. What did flowering plants do with their collections of newly duplicated genes?

The innovators did many things, united by a theme: Flowering plants turned "I" into "we." By intensifying relationships with other species, they made success a community affair. This is an old theme, long preceding the appearance of flowering plants or plants of any kind. The first cells on the planet, 3.5 billion years ago, lived in communities held together by reciprocity and mutual interdependence. But flowering plants took this ancient practice of symbiosis and extended it in unprecedented ways. Of all these floral mergers of lives, orchids exemplify the wondrous and sometimes troubling ways that close relationships can cause explosions of diversity.

Orchid

Intense relationships

ignite diversity

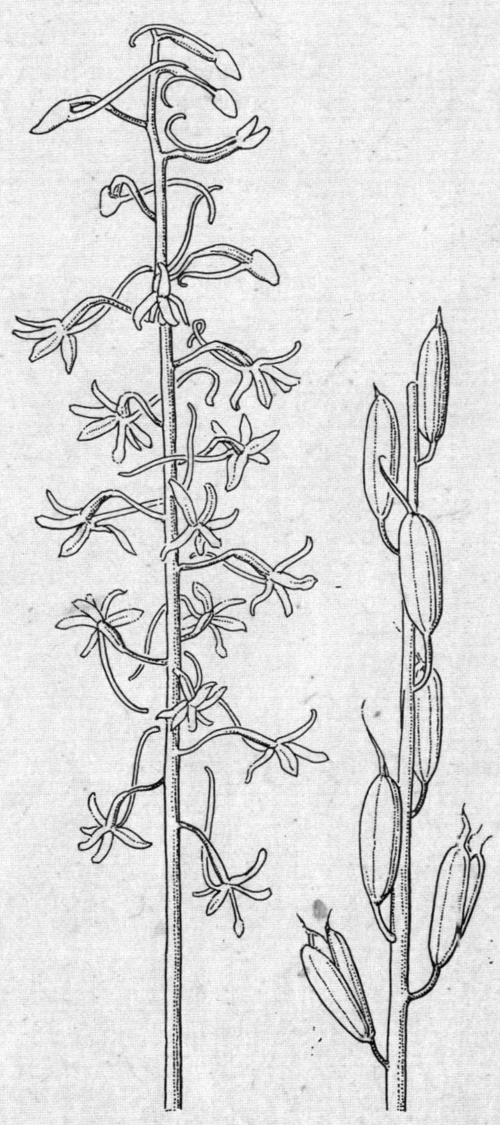

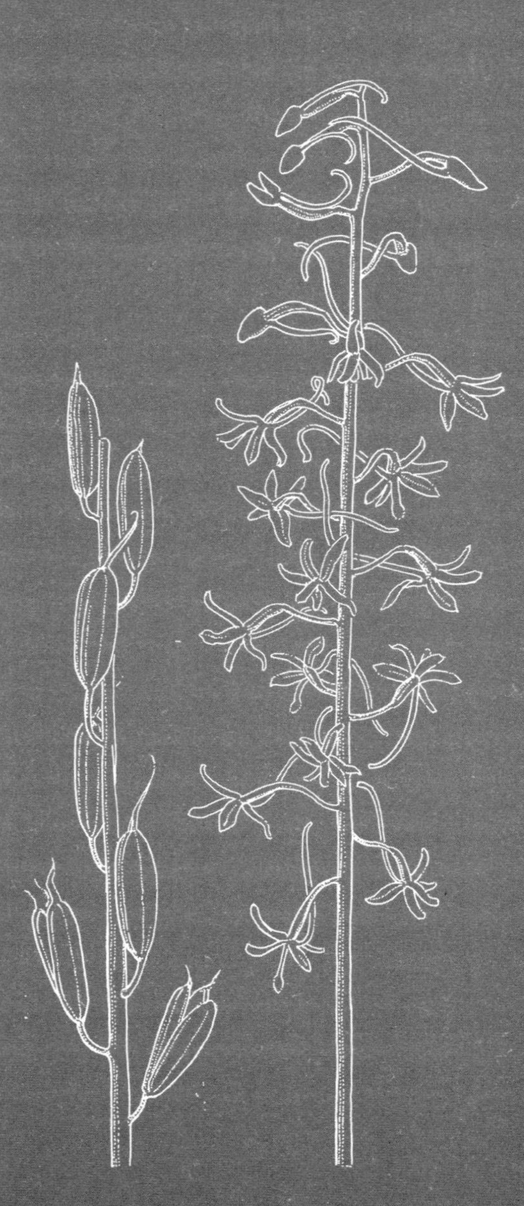

ODEPENDENCE LED TO SUCCESS. WHAT A CON-
trast to the dominant narrative of our culture, where individuality is lauded. "[I] earned my living by the labor of my hands
only," wrote Henry David Thoreau, echoed more than a century later
by British Prime Minister Margaret Thatcher, "Who is society? There
is no such thing! There are individual men and women and there are
families . . . society as such does not exist except as a concept," and a
United States presidential candidate, "I alone can fix it." Biological and
economic theory, too, is founded on the primacy of individuals, in
Charles Darwin's natural selection and Adam Smith's market economics.

This focus on individuality obscures the fact that individuals are created and sustained by relationships. Innovation and productivity, too,
are the fruits of interconnection. Boundaries between "I" and "we" are
impossible to draw. Without society—genes working together, cells
within bodies in constant conversation and exchange, energy flowing
through food webs, species interacting—there are no individuals. In this
matter, Thoreau was wrong: We earn our livings only and always
through the labor of many joined hands. The triumph of flowers is a culmination of this truth. Nowhere is this more evident than in the most
species-rich of all plant families, the orchids. Their alliances interweave
sky and soil, flying pollinators with root-bound microbes and fungi.
Many other flowering plants do the same, but the interdependence
achieved by orchids is especially intense. For many orchids, dependence
on others is so profound that it seems at first impossible that the plants
could survive or breed, let alone multiply.

Orchids first evolved 112 million years ago, and the family now contains about twenty-eight thousand species, making the orchids the most species-diverse plant family on the planet, although the aster and sunflower family is a close second or, by some counts, slightly more diverse. For comparison, there are fewer than seven thousand species of mammal. Each orchid species is a unique twine of plant, pollinator, and fungus. It is this closeness of relationship that fuels the orchids' diversity. In many species, the bonds are mutualistic. Almost all newly germinated orchids, for example, are entirely dependent on fungi for their food, a radical departure from the sunlight-fueled growth of most young plants. Orchid pollination is often highly specialized, using miniaturized and intricate sexual parts to work with a handful of pollinating insects or sometimes a single pollen-ferrying species. It is tempting to find in these connections an allegory about the harmony and generous reciprocity of nature. But other orchid species undermine the parable. Relationships take many forms. Some orchids have embraced illusion and thievery with gusto, parasitizing fungi and duping pollinators. But whether parasitic or mutualistic, orchid relationships are always intense. They have taken novelist E. M. Forster's exhortation "Only connect!" to extreme and sometimes twisted ends.

❧ ❧ ❧ ❧ ❧

As autumn progresses and most plants in the forests of the eastern United States fall into slumber, the cranefly orchid wakes. As falling tree leaves presage winter, we get an unexpected glimpse of new life. A pointy-tipped oval leaf pushes through the downed oak and maple leaves. The orchid's leaf is pleated and its top surface is a dark matte green, the color and size of a spinach leaf. Bend down and flip the leaf and you'll get a surprise. The underside is uniformly purple, sometimes as light as lavender but usually dark plum. The purple pigment salves the burn of frost inside the leaf. In a few individuals, the top of the leaf is spotted with purple. In others, the whole leaf is dark eggplant, one of the most striking and beautiful winter sights in these forests.

Autumn seems an odd time to unfurl a leaf, but the summer tree canopy here is thick. At midsummer, only meager greenish light trickles through. Many wildflowers emerge in early spring, then rush through their life cycles before the tree buds open. Cranefly orchid is hardier. It grows through the winter, then dies back as the spring wildflowers reach their peak. With little competition, the orchid leaf feasts whenever the sun is out and the temperatures edge above freezing.

The cranefly orchid rests all its hopes on one leaf. This precious solar panel is well defended. Microscopic spikes of calcium crystals are scattered throughout. Purple pigment makes the leaf hard for deer to see. As it sips on weak sunlight, the leaf sends food to a buried stem. In late spring, the leaf dies then quickly rots. Hidden in the soil, the stem waits out early summer, its tissues swollen with starch.

Sometime after the summer solstice, usually in July, a leafless stalk lances up. Straw thin, but as high as my calf, the stalk is first tinged with green and purple, then matures to the color of old hay. Flowers—sometimes two dozen or more—grow from the sides of the stalk, regularly spaced all along the top half. They look like half-crushed leggy insects hanging in a spider's web. Seen through a magnifying lens, the flowers sprout seven brownish-yellow skinny petallike projections and a long, slender nectar tube. The projections are twisted and squashed to one side. The overall effect is like a wounded cranefly, its legs akimbo, hence the orchid's name.

Against a backdrop of summer foliage, the flower stalks are easy to miss. To find them, it is easiest to note where the winter leaf grows, then return in summer to find the bloom. The flower might be inconspicuous to us, but not to owlet moths. For these small dusk- and night-flying moths, the blooms are an aromatic delight. A delight, that is, until the flower partly blinds them.

At night, the cranefly orchid bloom emits an aroma like a mild daffodil flower, so gentle that it is hard for my nose to distinguish from the smell of the dead leaves on the forest floor. When a cruising owlet moth picks up the scent, though, the signal comes through loud and clear. The moth flutters down and uncoils a long proboscis. Ah, sweetness. The throat of the flower leads to the backward-pointing spur full of nectar. The moth pushes forward, slurping deeper. Bump! One of the moth's eyes hits something very sticky. Orchid superglue gloms the eye. The moth pulls away and a pair of pollen sacs pop loose from their hiding place in the flower's roof. The pollen sacs are cemented to one of the moth's eyes.

The surfaces of moth eyes are smooth and hard, good places for glue, unlike the fuzz of scales that covers the rest of their bodies. The cranefly orchid's strange asymmetry, petals twisted to the side, evolved to hit the eye square-on. The arrangement works, at least from the orchid's perspective, because one-eyed moths keep searching for flowers.

Pollen sacs are first covered by a prophylactic cap that falls off after about thirty minutes. The cap prevents self-fertilization. Such "selfing" yields inbred offspring, although many plants, including the cranefly orchid, will eventually self-fertilize if they can find no other pollen. Once the cap is off and the moth finds another flower, pollen sacs get stuck onto a sticky stigma as the moth drinks. Half of the flowers twist left and half twist right. When moths visit a second flower, the pollen sac has a fifty-fifty chance of finding a stigma with the same twist.

The moths' love of orchid blooms partly blinds them. Photographer Ken Childs has recorded on his farm in Tennessee moths with half a dozen or more pollen sacs glued to each eye. When I've poked a toothpick into orchid flowers—including a few cranefly orchids—I've found the sticky pollen-delivering pad extraordinarily annoying. It has a gummy texture and will not let go. I try wiping it away with another toothpick and both get gooped, even after several hours. The orchid is unquestionably harming its pollinator by dimming moth vision.

Although the cranefly orchid stalk can hold two or three dozen flowers, on average only about seven mature into fruit pods. These swell through the late summer until they are bean sized. The energy and material needed to grow these fruits depletes the plants' underground stores. For every fruit grown in late summer, the leaf that appears in autumn is 2 percent smaller. Heavy fruiting causes the plant to skip flowering for a year.

In autumn, the stalk and fruits dry out, turning the light tan color of bleached fallen oak leaves, a great match for the leaf litter. On dry days, the fruit pods split, revealing shockingly small seeds. Through a magnifying lens I count thousands in a single pod. Like nearly all other orchid species, the cranefly orchid takes a minimalist approach to childcare. Each seed contains an embryo and nothing else. No droplets of nutritious oil, no starchy reserves, no protein packs. Just a ball of embryonic cells enclosed in a minuscule balloon-like seed coat. The gentlest touch of wind carries these mini zeppelins aloft.

When the flyers land, only those that alight on fungus-filled downed branches or tree stumps will live. Like all other orchids, germinating cranefly orchids draw their food not from the sun, but from fungi. The orchid babies welcome webs of fungal strands into their tiny bodies, then nurse on this inner network. Every orchid species has its own taste in fungus helpers, and cranefly orchids need wood-digesting fungi, usually found in abundance on fallen trees. In this way, parts of giant fallen trees are, via fungi, resurrected as minuscule orchids. This dependence of orchid on fungus can last for many years. When the orchids finally get large enough to grow leaves, they may repay the fungi with sugars, although by then the original nurses may have been replaced in the orchids' tissues by other fungus species.

Cranefly orchids are living headstones. Wherever we encounter them, we know we're at the gravesite of a fallen tree or branch. The downed wood may be gone, but the orchids let us peer back in time and know that we're in a healthy forest, one with a history of life-giving rot. Decaying wood provides food and sustenance to many: orchids, fungi, salamanders, small rodents, and many invertebrate animals. A forest newly grown back from pasture or one intensively managed for timber lacks a history of fallen wood. Cranefly orchids reveal this past.

The curious life cycle of the cranefly orchid is built, in all its parts, on specialization. This is true of all orchids. They thrive by focusing. Orchid seeds are as small as a plant seed can be. But, with the help of fungi, orchid seeds explore wherever the wind takes them. Orchid flowers are also stripped down. Although their petals are often showy, usually with a distinctive swollen lower lip, the female and male parts of the flower are miniaturized and fused into a central column. The pollen of most orchids, like the cranefly orchid, is bundled into two compact sacs, delivered to insects on a sticky arm.

Orchids thus take the bisexuality of ancestral early flowers to the next level, shedding and receiving pollen from the same compact structure. Their sleek flowers waste no pollen. In addition, the column-inside-petals design allows evolution to sculpt each orchid species' flower to the shape and aesthetic tastes of a small number of insect species. This allows orchids to develop specialized, tight relationships with insects.

Orchids are the anti-magnolias. Magnolias offer generous all-you-can-eat pollen buffets from open cups to any passing animal. Orchids invite only VIP insects to their blooms, often a single species of insect, then entrust these visitors with small parcels of elegantly wrapped pollen.

*　*　*　*　*　*

What do pollinators and fungal partners get from orchids? The plants thrive on these relationships, but benefits for their partners are often ambiguous, at best. Are orchids scam artists?

Consider first the fungi. Every orchid depends on fungal partners, especially when the orchids are young. Orchid seeds have no food stores of their own and grow into seedlings with no green tissue. Rather than feeding on sunlight, as almost all plant seedlings do, orchid embryos turn into miniature potato-like structures that send welcoming chemical signals to fungi. Fungal strands weave between the orchids' cells, then slither inside, where they coil and knot. Orchids digest the knots, sucking up fungal nutrients. Even though they are getting eaten, fungi keep growing, worming new strands into the orchids' hungry bodies. The diet of young orchids is therefore like that of a fungus-grazing animal. Although many types of plants merge their root cells with fungi, the knot-and-get-eaten pattern evolved only in the ancestors of orchids.

Orchids are generally selective about which fungi they allow inside, more so than many other flowering plants. In some orchid species, only a handful of genetic variants of a single fungus species are allowed in. In others, just one or two species of fungus are permitted. A few orchid species welcome numerous fungal types, but even these few are a small minority of the thousands of possible fungi that live in the soil and tree branches where orchid seeds land. In the controlled conditions of the laboratory, orchids will merge with many different fungi, but in woods and meadows they focus.

From bloom to root, orchids are picky about relationships. If you're going to be entirely dependent on another species for your germination or pollination, best be careful. If these relationships make you vulnerable, all the more reason. A young orchid that lets fungal strands into its tissues could get eaten alive by the wrong fungus. A flower with just two pollen sacs cannot afford to have them robbed.

Selectivity also hints that the orchids might be thieving. Parasites and pathogens are almost always specialists. Human head lice infect us and no other species, not even our close primate cousins, who must contend with their own louse species. Of the two hundred species of blood parasites that cause malaria, only five infect humans, four of which have specialized on our species. Disease-causing species sometimes jump among species—human tuberculosis, came from livestock, some human flus from birds—but these leaps are relatively rare. To succeed, a thief must perfect a narrow range of skills. Seldom can parasite species thwart every immune system or find a way to successfully transit themselves among many host species. Could orchids be parasites? In their relationships with fungi, the answer seems to be yes, at least much of the time. Orchids, in this view, are the head lice of the fungal world.

The evidence for parasitism is strongest in the youngest stages of the orchids' lives. Fungi get nothing obvious from the orchids that feed on them. Why would a fungus put up with such exploitation? Evolution has endowed fungi with great enthusiasm for growing into plant tissues. Some digest dead wood. Others form cooperative relationships with plants, exchanging minerals for sugars or oils. If a young orchid gives off the right signals, it can invite fungal strands into its body, then exploit the guests. This only works if the orchid's welcome is tailored to the tastes of its victim, hence the specialized nature of the con.

The orchids' small size helps them to succeed as parasites. A germinating seed is minuscule compared with a dead log or a living tree. Like a supermarket that tolerates a small amount of shoplifting to avoid security measures that would slow workers and inconvenience honest shoppers, fungi may simply put up with light-fingered orchids. This is partly conjecture. Fungi might also gain hidden benefits, such as a safe, moist place in which to ride out droughts or hide from pathogens. But fungi are such hard-to-study creatures—their bodies ever-shifting networks of almost invisible strands associated with other fungi and plants in mysterious ways—that precise measurement of the evolutionary costs or benefits of their interweaving with orchids has so far been impossible.

Seemingly parasitic life continues into adulthood for many orchids. About 250 species appear to be fully parasitic, gaining all their sustenance from fungi. This amounts to about 1 percent of all orchids, a larger proportion of parasitic species than in any other plant family. The bird's-nest orchid, for example, lives in woodlands across Europe and North Africa. It siphons some of the flow of food among beeches, hazels, and oaks. The orchid grows a "bird's nest" of roots that plug into the fungal network that interconnects these trees. By drawing its meals out of the fungus, the orchid indirectly parasitizes the trees. Like other entirely parasitic orchids, the bird's-nest orchid has no chlorophyll. When it pokes its head aboveground with a leafless flowering stalk, its tissues are a bleached beige color.

Fully parasitic orchids are an extreme version of what all orchids experience as youngsters. But even in orchid species with lush green adult foliage, the fungus-plant relationship is often heavily tipped in favor of the orchid. Among studies of orchids, there are many examples of food flowing from fungus to plant, and few showing the reverse.

The potted orchids in the supermarket or garden center give us a misleading idea. These plants seem to be solitary individuals, self-sufficient if given enough light, fertilizer, and water. But this is an illusion. Humans substitute for fungi. Unless we're buying an expensive and rare plant, horticultural orchids were grown not from seed, but from cuttings or cells grown on agar. These clones were then drip-fed the food they needed, food that, in the wild, would be provided by fungi. But even these progenies of labs and greenhouses have roots full of fungi, presumably picked up from spores in the air or in the growing medium. One study of greenhouse-grown *Phalaenopsis*, moth orchids commonly sold as houseplants, found over two hundred genera of fungi in their roots. I look with new eyes on the half dozen on our windowsill, most rescued from the "on sale" shelf and nurtured back to bloom. Each one is a living community, carrying in its roots the partners that make orchid life possible. As they face the constraints of living indoors, I'm a helpful addition, a clumsy pseudo-fungus carrying a mineral-rich watering can. Even when we grow orchids from sterile petri dishes, they find ways to draw others into relationships.

Utter dependence on another species for germination and establishment is risky. Orchid seeds in the wild must land not just on the right fungus, but also arrive when conditions are right for fungus and germinating embryo to grow and meet. Long odds. But orchids are successful gamblers because they make so many seeds, each one the size of a mote of dust. Four million seeds are enclosed in a single pea-pod-sized fruiting capsule of the green-lipped *Cycnoches*, an orchid from South American tropical forests. Most orchids spawn tens or hundreds of thousands of seeds, often from plants small enough to fit in your palm. This approach to parenthood makes orchids like ocean-dwelling animals. Corals and clams loose thousands of tiny, poorly provisioned embryos to the currents, hoping that a few find good landing spots, just as orchids cast dust seeds to the wind.

In his book on orchids, Charles Darwin calculated that the "great-grandchildren" a single orchid would, if every seed grew to adulthood, "clothe with one uniform green carpet the entire surface of the land throughout the globe." Impressive as this is, he thought the plants' "almost infinite number of seeds" was a sign of "lowness of organization" and "poverty of contrivance." After all, most other flowering plants generously provision their departing youngsters. We humans depend on these maternal gifts. The oils, starch, and protein in agricultural grains were originally intended to feed rice and wheat seedlings, not us. In contrast to the rest of the flowering plants, orchids seem tightfisted in their refusal to help the next generation.

But Darwin was mistaken. Far from being "low" in the evolutionary story, orchid seeds and seedlings take symbiosis to extraordinary and highly successful lengths. They do this partly by becoming brigands, some working in the dark back alleys of the soil, others threading themselves into bark and duff on branches in lush tropical treetops. Thus rooted, they extend flowers to pollinators. There, relationships are even more fraught.

❧ ❧ ❧ ❧ ❧ ❧

The most well-thumbed book in my childhood home and in my mother's house today is the 1978 edition of *A Field Guide to the Orchids of Britain and Europe*, by John and Andrew Williams, illustrated with lush but precise watercolors of each species by Norman Arlott. On the inner cover, my mother inked her name and the date she bought the book, "Jean Haskell, April 1979," but the most striking annotations are the penciled lists of species that crowd the inside covers and extend to inner pages. The lists record dates and places of her orchid finds, some on trips to meadows or woods and others close to home. My mother's love of orchids was kindled by undergraduate cycling trips in the Chiltern Hills near Reading, England, organized by the Bot Soc, the university club for botanizers. When she moved to France as a young woman, she was stunned to see species that she knew as rarities in England growing in profusion in French verges and fields. Not only was the more southerly climate friendlier to orchids, but in the 1970s much of the French countryside had yet to be impoverished by industrial agriculture. On some weekends in early summer, our family drove out of the city to search for orchid-rich meadows.

The site that stays most in my memory is on a curve of the river Seine, west of Paris. There, the river has sliced through limestone, leaving high cliffs on the outer curve. Such natural fortifications made the area an important military stronghold and crossing point. A ruined twelfth-century castle and an eighteenth-century château perch in the small town of La Roche-Guyon. We headed for the higher slopes, along one of the "sentiers de grande randonnée," long-distance hiking trails that thread through much of western Europe. Our walks were relatively short, but for my sister and me, these were grand adventures. Looking at a satellite image of the area now, I see that the uncultivated area is a narrow strip along the high limestone bluffs, with agricultural fields and managed forests behind. But for kids growing up in the city and suburbs, the place felt spacious and wild. We grumbled about what we thought was the slowness of orchid searching, but reveled in the unmowed grassy glades between patches of open woodlands. In those days, the 1970s, we could not step onto the grass in Parisian parks without getting scolded by park guards who kept kids in line with shrill metal whistles. In the city, green was off-limits. Not so at what we called "the orchid place."

The strange names of the orchids near La Roche-Guyon enchanted me then and still do. Like characters from mythology, they have a chimeric quality, merging plant and animal: lizard orchid, monkey orchid, lady orchid, man orchid, early and late spider orchids, bee orchid, bird's-nest orchid, and military orchid. Other names evoke shape and aroma: twayblade, fragrant, and pyramidal orchids.

My favorite is the fly orchid. The name makes my skin crawl a bit, a reminder of annoying insects. The flower looks like a fat-bottomed insect hovering in front of a green-yellow bloom. The "insect," made of petals, is a beautiful velvety reddish-black, marked across its midline with a silvery band that shines like insect wings. Two winglike extensions poke out on either side and, best of all, long antennae curve from the head. The scientific name, *Ophrys insectifera*, means insect-like eyebrow orchid (*ophrys* refers to brow-like hairs on the flowers). I learned that the fly orchid is a sexual illusionist, casting lusty spells on insects. Case closed. For a kid, flowers that conjure mirages were fascinating. The same is true today. Beautiful faces with edgy, complicated backstories are alluring.

The fly orchid aims to deceive. Like the flower wielded by Puck in Shakespeare's *A Midsummer Night's Dream*, the orchid casts a spell that induces misplaced desire. Through sensory illusion, the fly orchid dupes male wasps. Like Titania, the wasps become enamored with the wrong species, in this case the orchid flower itself. Most other orchid species are named for coincidental resemblance, like finding shapes in clouds. The monkey orchid has four arms and a tail, the spider orchid sports legs and a fuzzed abdomen, and the military orchid is helmeted. But the fly orchid flower looks like a female wasp (or, to human eyes, like a fly, hence the name), a mannequin sculpted by evolution to besot male wasps. When male wasps wake from their underground winter slumber, they emerge into meadows strewn with what seem to be lovers. *What angel wakes me from my flow'ry bed?*

Looks are stimulating, but aroma really gets male wasps going. And so the fly orchid mimics the smell of female wasps. Not just any wasp—there are hundreds of wasp species across the fly orchid's range—but the females of a black-and-yellow-banded wasp, *Argogorytes mystaceus*, a species common across Europe and North Africa. Females build underground nests, usually in bare soil, and provision the nest chambers with paralyzed spittlebugs and other small insects on which the young wasps feed. Males contribute nothing to breeding but sperm. The orchid flower wafts wasp love-perfume in the air. The mimicry is precise, matching the blend of chemicals used by females. Orchids make such good knockoff perfumes that their effects can be more exciting to male wasps than the real thing.

Smelling what he thinks is a chemical come-on from a female, the male wasp zips toward her. Seeing the shape of a female, he lands on her back. *On the first view to say, to swear, I love thee.* There, the fuzzy abdominal hairs further stimulate him. Before the illusion breaks, the orchid glues stalked pollen sacs to his back. If the hapless suitor is fooled again, pollen reaches the female stigma of another fly orchid. But wasps are clever. They remember the flower that fooled them. That smell, though! So enticing. *How I dote on thee!* Another flower often tempts them. The fly orchid's pollination strategy walks a fine line between the salaciousness and prudence of male wasps.

What female wasps make of their mates' floral dalliances is unknown. As long as they have a selection of males from which to choose, the orchids' interference may be of little consequence. But males occasionally waste sperm during their orchid encounters, making them less valuable mating partners. This happens rarely for wasps visiting *Ophrys* orchids but more commonly for wasps visiting the deceptive tongue orchids of Australia. Males presumably also pick up scent as they embrace the orchid. This could show that they are vigorous, active fliers. But it could also falsely signal that they have already mated with another female and are thus depleted of sperm and potentially carrying disease. The curious genetics of wasps and bees adds another twist to the female wasps' perspective. For wasps, bees, and ants, unfertilized eggs grow into males. Fertilized eggs become females. Female wasps, like some of their kin the bees and ants, can therefore breed without mating, albeit producing only male offspring. Because females can, in a pinch, get along without mates, these insect species may be more vulnerable to sexual deception by orchids.

Fly orchids are hardly alone in bending the desires of male insects to their own ends. Of the thirteen orchid species that we found at La Roche-Guyon, four solicit male insects. The early spider orchid dupes miner bees. The bee orchid specializes on long-horned bees. The late spider orchid attracts chafer beetles and other insects. Other members of the bee orchid's genus, *Ophrys*, beckon to bees, wasps, and flies. Each orchid species makes an aromatic blend specific to just one, or sometimes a handful, of insect species.

Ophrys is just one example of orchids' penchant for sexual deception. Worldwide, about one sixth of orchid species use some form of sexual mimicry to attract pollinators, mostly wasps and bees. The most bizarre are perhaps the hammer orchids of Western Australia. A side arm from their flower stalk holds what looks and smells just like a female wasp. When male wasps grasp and try to fly away with their beloved, the side arm swings up. The flower slams the amorous male into a cup containing glue-covered pollen sacs.

Sex is not the only illusion conjured by orchids. Lying about food is even more common. These flowers promise nectar or pollen, but deliver none. Five of the thirteen species at La Roche-Guyon are "food deceptive." Worldwide, about 40 percent of all orchid species offer no apparent nectar or other reward to visitors. There's a nice flower, a passing insect thinks. Why not visit? Once in the flower, they find no edible treats.

Many of these foodless orchids have reshaped their flowers to offer false advertisements of nectar or pollen. Some grow empty nectar spurs. Others mimic pollen, splashing their flowers with yellow pigment or dangling fake anthers. To better fool insect eyes, the pollen-mimicking parts of the flowers absorb ultraviolet light, just as real pollen does. *Maxillaria* orchids from South America go one step further, strewing the entrance to their flowers with pollen-like dust grown from tiny hairs. In some species, this imitation pollen is provisioned with protein and starch, giving insects a meal. No con there. But in other species, the dust is missing protein, starch, or both. These species are double mimics: Their foodless dust imitates the nutritious dust of other species, which, in turn, are mimics of real pollen, which is even more nutritious. Trust the orchids to make things complicated. In all these species, the flower's real pollen is hidden in small sacs that get glued onto hungry but disappointed pollinators.

A few orchid species mimic flowers of unrelated species. Some sun orchid species in Australia look like yellow pea flowers. The peas make nectar, but the orchids do not. Other sun orchid species mimic blue lilies, adding a fake anther for good measure. In South Africa, *Disa pulchra* orchids have almost exactly the same color as local *Campanula* bellflowers. Like the Australian pea-flower mimics, these orchids have reshaped their flowers to look more like their "models," the bellflowers. In all these cases, orchids parasitize the generosity of other plants.

Orchids are not the only plants to proffer empty promises of food. Dayflowers, the blue-flowered herbs that grow across much of the Northern Hemisphere, have large fake anthers that glow yellow against the blue flower. The actual pollen-producing anthers are smaller and sit below, ready to dab visitors. Many other flowers, including globeflowers and bog stars, do the same, waving at insects with visually impressive but empty anthers. The most well-known flower of this kind is the saffron crocus. The saffron "threads" that we use in cooking are anther mimics. The threads are reddish and rough textured and, to insects, look like pollen-loaded anthers. But when insects land, they find a pollenless stigma, the female part of the flower. Only by walking down the stigma, depositing pollen as they go, do the insects find real pollen. The golden hue that saffron imparts to our dishes is a pigment evolved to fool insect eyes.

In other plant species, female flowers mimic males. Squashes grow separate female and male flowers, perhaps to avoid self-pollination. Lacking pollen, the female flowers risk being ignored by insects. Squash plants solve this problem through deception. Their female flowers grow stigmas that look like pollen-shedding anthers.

The squashes' trick fools many insects, especially small solitary bees, but not honeybees. The honeybees quickly learn to discriminate male from female—the visual mimicry is far from perfect—and visit only male flowers. From the squash plants' perspective, these clever honeybees are herbivores, sapping the plants' energy and giving little in return. The nimble mind of the honeybee can distinguish among thousands of flower aromas and shapes, a capacity likely evolved in part to outsmart floral scams. In response to sophisticated bee learning, a few plants have upped the ante. The female flowers of climbing roses and wild relatives of tomatoes, for example, grow such convincing fake anthers, complete with imitation pollen, that even honeybees repeatedly fall for the visual illusion.

Many flowers spike their nectar with chemicals that manipulate pollinator behavior. Citrus and coffee plants, for example, give their pollinators a shot of caffeine mixed in the nectar. A caffeine buzz improves bee learning, making them up to three times more likely to revisit flowers that smell the same as the barista flower. At first glance, this seems to benefit both bee and flower. But, although the flower gets better pollination, the bees get hooked and stick with the caffeinated plants long after they would otherwise have switched to other, more profitable flowers. The flower has cast a spell on the bees' memories, creating an illusion of ample reward. The plants use just enough caffeine to stimulate the bees' nerves, but not enough for the bees to be put off by the bitter taste. The chemical relationship between flowers and insects is not all manipulative. Nectar sometimes also has traces of chemicals that kill the mites and other parasites that plague the bodies of many pollinators. Nectar is not just sugar water, it is a drug cocktail.

A few pollinators pay for floral deception with their lives. Jack-in-the-pulpit, a common wildflower in the woodlands of eastern North America, flowers by poking from the leaf litter a "pulpit," a tall, vaselike structure capped with an overhang. Inside stands the "jack," a vertical column covered on its lower surfaces with tiny flowers. Individual jack-in-the-pulpits can grow either male or female flowers, or both, usually switching from male to female blooms when the plants have accumulated ample belowground food stores to supply the fruits. The flowers mimic the smell of decaying mushrooms. This whiff of rot draws fungus gnats who are deceived into thinking that a dying mushroom is ready to receive their eggs. Once inside the pulpit, the gnats get caught in slippery wax and slide helplessly down the walls. If the insects fly into a male flower, they get dusted with pollen at the base of the jack and eventually wiggle out through a tiny exit hole. But female flowers have no exit. They are floral femmes fatales. Trapped inside the female flowers, the gnats deliver their pollen, then die. Relatives of jack-in-the-pulpit, plants in the arum family, have similar strategies. Some use catch-and-release, letting their gnats go after the insects to deliver the pollen. Others smell of dead mammalian flesh, attracting carrion flies and beetles. Because the flower contains no actual flesh, eggs laid by these pollinating insects are doomed.

For gnats killed by a jack-in-the-pulpit or insects wasting time and energy on "pollen" that turns out to be mere yellow sawdust, illusions are costly, clear examples of plants harming their pollinators. Such deceptions work only if the orchids are relatively rare. Do we stop eating apples because occasionally one has a rotten core? No, but if every apple were nasty, we'd learn not to bite into them. If the world had no flowers with real pollen and abundant nectar, or no real rotting mushrooms, food mimicry would not be possible. Sexual and food deception only work because insect senses also often faithfully guide them to mates and food. A fungus gnat dying in an inescapable jack-in-the-pulpit is collateral damage from evolution's calculus. As long as most gnats find food and breeding sites, and not death, natural selection will give the insects a keen interest in the smell of rotting mushrooms.

Orchids deliciously complicate the human urge to read allegories into "nature." We are often quick to find morality tales in the lives of other species, either benevolent cooperation and kinship or ruthless free-market competition and exploitation. There can be truth and wisdom in these stories, especially if they are grounded in generations of thoughtful study, but the tales often reveal as much about our beliefs and aspirations as they do about the creatures whose lives we interpret. We project our values onto flowers, forests, or animal behaviors, then claim that "nature" is offering us a lesson. Orchids tell us: not so fast. They live in total dependence on fungi and pollinators. These deep relationships take many forms, from reciprocity to piracy. We cannot spin a simplifying moral conclusion from orchid lives. Instead, we might conclude from orchids that, against long odds, it is possible to thrive in the tension between boundless generosity and craven exploitation. Often both are true, in the same densely knotted web of relationships that is an orchid.

✶   ✶   ✶   ✶   ✶   ✶

At La Roche-Guyon, my mother's orchid list runs to thirteen species. For a single plant family on a small patch of stony ground far away from the tropics, that's an impressive count. Some taxonomists estimate that *Ophrys* alone contains 354 species, although others put the number in the dozens. Worldwide, orchids are the most species-rich plant family, with asters and sunflowers close behind (or slightly ahead, depending on exactly how you define a "species"). Taxonomic "families" are unstandardized human delineations, and many species have yet to be formally named, but the statistic reminds us that the orchid way of life has been phenomenally successful at producing new species. This evolutionary creativity grows from the relationships that orchids forge with other species. The space between ecological generosity and exploitation is tense, but fruitful. The closest and weirdest relationships are often the most generative.

Sexually deceptive *Ophrys* flowers show us how specialization leads to diversity. The male insects that pollinate *Ophrys* flowers only pay attention to specific blends of scent, the particular aroma of females of their own species. When these males get duped by orchids, this specificity makes them faithful couriers of pollen, always visiting the same orchid species and ignoring all other flowers—these others just don't smell right. Orchids of different species can therefore grow right next to one another and never exchange pollen. This is unusual. Most flowers are visited by insects that carry pollen from many flower species. The female stigma of these flowers can "taste" pollen, using chemical receptors, and control which pollen grains germinate and grow. When pollen of the wrong species arrives, it will usually fail to develop or the embryo will be inviable. In *Ophrys*, no such barriers exist to the growth of pollen or the development of embryos. Instead, insect sensory preferences alone maintain separation between *Ophrys* species. This total reliance on the senses of picky insects creates a powder keg for evolution.

The aromatic signature of *Ophrys* flowers is largely shaped by one or two genes. A single mutation can produce a new aroma. When this happens, the original pollinator insect loses interest. The flowers of the mutant plant will usually be loveless, their changed smell failing to attract insects. But occasionally a novel aroma will summon a new lover, a different species of insect. This pollinator switch can create a novel species of orchid, one pollinated by newly recruited males of an entirely different insect species. This process seems to have happened dozens or hundreds of times in the *Ophrys* genus.

Genetic studies show that *Ophrys* is a relatively young genus. The living species date their ancestry back about 5 million years. In this, they are like many other orchid species. Although the orchid clan is at least 112 million years old, most of the diversity among living species arose in the last 5 million years, perhaps driven by fragmentation of habitats as global temperatures cooled. Judging from the orchid family tree, ancestral *Ophrys* plants were pollinated by wasps, as the fly orchid is today. As a cooler climate turned lush Eurasian and African laurel forests into grassland and scrub habitats, *Ophrys* evolved new species that beckoned to long-horned bees. Then, in just the last million years, the rate of speciation surged as *Ophrys* latched onto yet more bee species. Today, these new *Ophrys* species attract males of miner bees, flower bees, plasterer bees, mason bees, leafcutter bees, carpenter bees, and cuckoo bees. This burst of diversification got help from "jumping genes," bits of DNA that copy themselves and cause parts of the genome to multiply. It seems that these jumping genes had a spasm of activity about one million years ago, facilitating the multiplication of hundreds of genes, most of which affect flower development, color, and aroma. As is often true in flowering plant evolution, genetic duplications and rearrangements catalyzed novel and diverse ecological relationships. Or, put another way, ecological opportunity allowed some genetic revolutions to succeed.

Looking further back in the orchid family tree, we see that sexually deceptive orchids emerged from ancestral species that lied to insects about food. Just as many orchid species do today, these ancestors promised nectar and easy-to-access pollen but provided none. From this common sleight of hand evolved the more elaborate aromatic, visual, and textural deceptions of *Ophrys* and other sexual mimics. "Oh, what a tangled web we weave, / When first we practice to deceive!" Walter Scott's poem about human romantic deceptions applies to orchid evolution, too.

The fly orchid and its kin are an extreme example of a more general feature of flowering plant evolution. Close bonds between plants, insects, and fungi often lead to increased speciation. Relationship begets diversity. Deception is not required. The Eurasian and North African lesser butterfly-orchid, for example, provides ample nectar from its spike of green-white flowers for hawkmoth pollinators. Not all moth species have the same length of proboscis, though. Those in woodland sport longer proboscises than those in meadows. The orchid has evolved different lengths of nectar spur to match these differences. For now, there is enough cross-pollination between woods and meadows to unite the orchid's gene pool. But if the ecology or behavior of the moths change, the orchid species could split into two.

Pollinator switches also cause rapid evolutionary change in other flowering plants. Lab-grown mustard plants pollinated by either bumblebees or hoverflies evolve their flowers to match the insects' preferences. After just eleven generations—a generation for some varieties of mustard is about forty days—the bumblebee-pollinated flowers are taller, more aromatic, use strikingly different aroma blends, and have larger ultraviolet reflecting areas. Each flower type attracts its own pollinator most effectively.

Flowers are not fixed entities. Through natural selection they shape themselves to the sensory systems of their pollinators. Good advertisers and illusionists understand their audience. Floral evolution in response to changed pollinators is a major driver of flowering plant diversity. Across all flowering plants, about a quarter of splits in the family tree are associated with switched pollinators.

It is not just pollination that drove the world-changing co-diversification of flowering plants and animals. To deter the hungry mouths of insects, flowering plants infuse their tissues with chemical feeding deterrents and poisons. Insects often evolve countermeasures, detoxifying the plants' arsenal. But to work, countermeasures must be chemically specific and so insect species tend to specialize on one or a few plant species. This tight back-and-forth, an evolutionary arms race, between plant and leaf-eating insects makes populations brittle. They often splinter. Each subpopulation evolves its own coupled set of toxins and purges, a rich seedbed for new species.

Some symbiotic fungi, too, became much more diverse after flowering plants evolved, although many of the main branches of the fungal family tree were well established in the dead wood and living roots of the ferny, coniferous habitats that preceded the appearance of flowering plants.

For orchids, other flowering plants were another spur to diversification. When modern tropical rainforests appeared, created by the prodigious growth and water-laden breath of flowering plants, orchids moved to the trees and repeatedly evolved new species. Today, the branches and trunks of many tropical trees are festooned with orchids that complete their life cycles in the tree canopy. These epiphytes ("plants that grow on others") account for 70 percent of all living species of orchid. Along with other flowering plants like bromeliads, half of all plant species in rainforests are epiphytes. These plants clinging to branches and tree trunks provide habitat for tens of thousands of animal species, from pollinators, to herbivores, to the microscopic animals that live in the moisture of leaf hollows and roots. If you have orchids in your home, it is likely that they grow not in soil as other houseplants do, but in moss-filled open containers that mimic their tropical treetop homes.

❦ ❦ ❦ ❦ ❦ ❦

As a child, I experienced La Roche-Guyon as a single place, an orchid-strewn limestone outcrop, immersed in the hum of bees, flies, and wasps. But from the perspective of orchids and insects, multiple sensory worlds coexist. Some are straightforward, others are built on illusion. The intersections among these worlds create and sustain the gorgeous diversity of plant and insect.

For the fly orchid and its *Ophrys* kin, the bee orchid and late and early spider orchids, the sensory tastes of single insect species are all that matter for pollination. The flowers of monkey, lizard, lady, and pyramidal orchids call out with ornate flowers to a diverse array of insects, but give no reward. The fragrant orchid offers long-tongued moths nectar from spurs at the back of the flower. Bird's-nest orchid smells musty and draws flies to its meager nectar. The other flowers here—brambles, hawthorn, buttercups, and cowslips—lack the duplicity of many orchids and use nectar and pollen to train insects to love flowers, a schooling that has bred floral enthusiasm into the insects' genes and nervous systems.

The insects have a different perspective, visiting many flowers to gather nectar and pollen, sometimes finding rewards, sometimes empty blooms. They remember individual flowers and aromas, shapes, and colors, and are guided by this past experience. A few, the moths and short-tongued bees, focus their efforts on the flowers that their mouthparts can access; others, hoverflies and some bees, are more catholic and buzz from one flower species to another.

For a few male insects, pollen and nectar are of little interest except for short refueling stops. Instead, for these males, sexual aromas slice through the crowd of other signals, guiding them either to a floral dummy or a mate. Many of these insects create illusions of their own. Wasps and bees use a shared visual language to speak to the eyes of predatory birds: bands of black and yellow mean danger from stingers, stay away from us. But the stingless male wasps and bees wear the warning, despite being harmless. Weaponless hoverflies, too, put on a deceptive cloak, painting themselves with danger signals and buzzing like honeybees.

Evolution spun this web of relationships. Some strands tightly link orchids and their pollinators; others stitch dozens of species into looser weaves. Every strand thrums with possibility for new arrangements, new diversity.

No other group of plants can match the extravagance of orchids for species diversity. Through deep relationship and extreme specialization, orchids embody Charles Darwin's famous closing words of *On the Origin of Species*: "From so simple a beginning endless forms most beautiful and most wonderful have been, and are being, evolved."

Orchids achieve their success partly by being rare. If your babies are parasites, there is a limit to how many can exist before their hosts reject them. If your pollinator is one species of insect, you can only live where they do. This scarcity drives some of the human fascination with orchids. Rare beauties are worth a drive out of the city to find. But even common orchids with wide ranges, like the cranefly orchid, are never abundant in any one place. In terms of biomass, they are dwarfed by most other plants. To find answers to how flowering plants became not just diverse, but so incredibly abundant that most ecosystems are founded on their green productivity, we need to look underfoot. As we marveled at the orchids, we were kneeling on grass.

Grass

All flesh is grass

All flesh is grass

G RASSES ARE THE CONTRARIANS OF THE FLOWER
world. For most of us, they barely register as flowers. Their
blooms have no gaudy pigment. They make no aroma, save for
the fresh tang of pollen. They offer no welcoming bowl of petals. Instead
grasses cluster small flowers along stalks, each flower shaking its pollen
from dangling anthers. Feather-like stigmas catch pollen and lead to an
egg buried close to the flower stalk. When the flower matures, it makes a
dry, papery fruit, with a starchy seed inside. Instead of launching upward and girding themselves with wood like trees, grasses crawl their
ever-skinny bodies along the ground.

Shunning conventional wisdom was a brilliant move. Grasslands, including grassy savannas, now cover about one third of the Earth's land
surface. This abundance is relatively recent. Ancestral grasses evolved
as early as eighty million to one hundred million years ago, when the
world was a hothouse covered with dense tropical forests and other lush
vegetation. They crept through the deep shade of the understory of
these forests and explored more well-lit areas on the forest margins. But
it is only in the last ten or so million years that grasses built vast prairies,
steppes, and savannas. Along the way, they also catalyzed the evolution
of early humans, and their bounty is our main source of food to this day.

Scientists have compared grasses to Vikings: tough adventurers who
disperse long distances to establish outposts, then permanent settle-
ments. The analogy captures grasses' wandering and tenacious spirit,
but omits other, less marauding qualities. Grasses built new body forms
and reimagined plant motherhood, provisioning their seeds so amply

that we now use them as cereal crops. Their wind-pollinated flowers do well in open, dry areas where insect partners may be unreliable. Suffer from hay fever? Thank the success of grass flowers. Grasses also evolved hidden biochemical superpowers, chemical wizardry inside their leaves. Last, grasses fully embraced multispecies ecological communities. Viking outposts turn into cosmopolitan cities, full of cooperative enterprise and innovation.

❧ ❧ ❧ ❧ ❧ ❧

"This area was all golf course. It was active twenty years ago. Look, you can see the remnants of where a sand trap used to be." Gabe Andrle points at a slight depression on the charred ground, light-colored sand contrasting with the reddish soil. "And the area we're standing on was used as a staging area when they reconstructed the dams. This part got the most disturbance of all."

We're at Panola Mountain State Park, just outside Atlanta, on a warm morning in March. A heavy dew weighs down spiderwebs in scattered clumps of briars. The humid air tastes earthy, with a hint of charcoal and wet hay. Gabe, a young man with a close-cropped beard, is the habitat program manager for Birds Georgia, a nonprofit that aims to build "places where birds and people thrive." Gabe's conversation overflows with excitement about birds, reptiles, and plant conservation. Today, he's showing me how setting fields on fire can help both grasses and birds.

Two weeks ago, Gabe and his colleagues torched this knoll and the meadows around it, an area of about five hectares (twelve acres). Fire-blackened trunks of young pine and callery pear trees crowd the woodland edges around the burned open area. A large oak stands at the meadow's center, seemingly unharmed. Underfoot, fine soot covers bare soil. We pick our way between remnant tufts of scorched grass and shrubbery.

As we start walking, it seems that fire has wiped the slate clean. But then I see green leafy shoots poking up from grass tufts.

"That's broomsedge. We've been working here for three years now and this area is really coming along. It had a slow start because of all the past disturbance, but this species really does well." We crouch to see the grass more clearly. The tufts are like stiff-bristled scrubbing brushes, prickly and tough. The bristles are last year's growth. The leaves emerging from them are soft and pliable, and their fresh blue-green contrasts strongly with the black and tan of the burned remnants. Looking more closely, I see brushes all across the meadow, new leaves poking from each one.

"We seeded broomsedge. It's our primary species on this site. But it also came in on its own. Here in the Piedmont, it tends to be the species that comes in first. When you're driving around Atlanta and you see a disturbed site, a construction site or whatever, broomsedge is what you typically see popping up." Not just Atlanta. If you've seen knee- or waist-high brown plumes on roadsides or old fields in the eastern United States, you've seen broomsedge. The plant is a true grass, not a sedge as its name implies. The tussock-like sedges look similar to grasses, but they belong to a different family and often grow in wet areas. Broomsedge thrives in challenging, often parched soils and its dried summer growth stands through the winter into spring and summer.

As we stroll, he points out some other native grasses, including a few that the flames skipped and left standing. "Here's little bluestem and split-beard bluestem; they got here on their own." The little bluestem is taller and redder than the splitbeard, with less puffy seeds. The colors are gorgeous, a mix of gold with red and purple tints. A haze of delicate seed spikes catches the light and glows silver. Lawn grass is as uniform and flat as a painted floor, but here, despite the fire, grasses grow in varied textures in dozens of tones and shades.

"There are also a few non-native species here, Bermuda grass left over from the golf course, and some Johnson and dallisgrass." Gabe waves dismissively at these tufts of unwanted grasses and explains that, over time, he and his colleagues in state agencies are using seeding, transplants, fire, and squirts of herbicide to steer the meadow toward native grasses, with some wildflowers mixed in. "I think some of the seed that we put out two and a half years ago is starting to express itself now." Gabe crouches again, grinning as he brushes his hand over the sprouted broomsedge. "Every site is different and pyrodiversity is part of that. We aim to burn here every year, but other sites get less fire." Without fire, this meadow would, in just a few years, turn into a thicket of blackberry, sweet gum, pine, and pear saplings.

Here at Panola, a mammal—in this case, humans organized into non-profits and government agencies—is working with fire to help grass thrive. This is an old story. Grasses and fire have been partners for tens of millions of years. Lately, some mammal species have also been part of the tight, mutually supportive relationship, mostly through grazing by large herbivores, but now also through fires deliberately lit by humans. The first grasses eked out a living in the shade of tropical trees, in the late Cretaceous. The fossil record shows that grasses first lived in the forest understory, then in small forest glades and open woodland, and more recently in habitats never before seen on Earth, open prairies, steppes, and savannas. Burning brought about many of these changes, helping grasses to expand their ranges whenever the climate turned seasonally arid or otherwise fire prone. We humans did not invent deforestation; grasses beat us to it, although on a slower timescale, one that created possibility for other plants and animals rather than erasing diversity as our bulldozers do today.

Broomsedge's resurgence just a fortnight after a fire reveals some of the grasses' special talents. Unlike most other flowering plants, grass stems grow sideways, not up. Grasses don't waste energy on woody trunks. Instead, they live either by burrowing underground or slithering along the surface, then lofting disposable leaves and flower stalks. Some grasses, like broomsedge, grow as "bunchgrass," keeping their horizontal stems vanishingly short so that roots and leaves grow from tufts. Others, like Bermuda grass, originally from Africa, have longer buried stems that wander along and through the ground, sprouting roots and leaves at regular intervals. Like broomsedge, Bermuda grass tolerates heat and drought, and patches of it persist in the old golf course despite repeated rounds of burning and seeding with native grasses.

Keeping low protects grasses' embryonic cells. In all plants, new growth comes from meristems. These are botanical fountains of eternal youth, clusters of undifferentiated fresh cells that can divide and transform into any of the mature cells needed in plant stems, roots, or flowers. Grasses keep their meristems buried, either in the ground or just above it in protective thatch. When fire comes, no problem. Trees and shrubs keep their meristems at the tip of every twig and in a layer under the bark, a wonderful way to expand into the aerial world, but vulnerable to fire. When this meadow burned, broomsedge got a haircut, but all the trees and shrubs, save for the thick-barked old oak, were killed. Now, after the fire, broomsedge meristems merrily push up fresh leaves to bask in unobstructed sunshine. As a bonus, they are fed by the fertilizing ash of their woody competitors. The grasses here were also spared the trees' lost investments in wood. By staying low and making cheap leaves, grasses are resilient and thrifty. The Israelite prophet Isaiah used grass as a symbol of ephemerality: "The grass withereth, and the flower thereof falleth away." For a few grasses that reseed themselves annually, he was right. But the seventeenth-century English writer and cleric John Donne saw a subtler truth about grasses. Most are defiantly perennial. They persist, even when their aboveground parts have withered and burned. He wrote that love "doth endure / Vicissitude, and season, as the grass," a theme echoed in the twentieth century by Max Ehrmann's *Desiderata*: "In the face of all aridity and disenchantment, it [love] is as perennial as the grass."

For most living creatures, fire is a calamity. Yet, paradoxically, plant life made fire. By the simple act of growing, plants invite conflagration by pumping oxygen into the atmosphere and filling the world with flammable wood and greenery. Ever since plants arrived on land, fire and plant life have existed in an uneasy and often destructive relationship. Grasses, and a few other fire specialists, flipped the nature of the relationship. For them, fire wasn't an inevitable disaster to be overcome, but a friend to be welcomed and even encouraged.

Grasses are not just passive beneficiaries of fire. They are the green Prometheus, domesticating fire for their own use by evolving to make fire more likely. They build the conditions for their own success, a form of evolutionary agency. When grasses are done with their leaves, they don't let them blow away or rot. Instead, in many species, dead grass leaves remain standing, resist decay, and ignite easily. The tussocks of dried broomsedge that stand all autumn and winter are incantations: *Come, fire, we've lived in preparation for you.* On the knoll at Panola, Gabe and his colleagues apply the match, but the fire was built by the grasses themselves.

In the later stages of grass evolution, starting about twenty million years ago, grazing mammals joined the grassland party. Just as the first flowers caused the evolution of pollinating bees and butterflies, grasses created grazing mammals. Mammalian teeth do much of the same work as fire. By clipping buds and stripping bark, grazers clear the understory, girdle mature trees, and exterminate shrubby seedlings. From the grasses' perspective, these grazing animals have the wonderfully useful quality of being self-perpetuating. By sacrificing a few leaves to the stomachs of mammals, grasses ensure that their landscaping crew will come back in future years. Another form of agency: The lawn grows its own lawn mowers and tree trimmers.

A diet of grass is hard on teeth. Grasses spike their leaves with tiny shards of glass made by special cells inside the leaf. Up to 5 percent of a mature leaf's weight can be silicate. This combines with grit from the soil to make eating grass like chewing on sandpaper. Grazing mammals therefore evolved tall chewing teeth, capped with sharp ridges of enamel. If ever you find a sheep or horse skull, examine a back tooth. They can be as long as your hand. Run your fingers over the teeth and feel how the top edges are as sharp and angular as broken flint. This dental innovation allowed mammals that eat grass—either exclusively or as a large part of their diets—to diversify and eventually dominate the large mammal community on most continents.

In the grassland near Atlanta where Gabe and I are exploring broomsedge, most of the large grazers that were common in this part of North America for hundreds of thousands of years are gone. Grasslands in the southeastern United States were home to grazers such as Columbian mammoths, Harlan's ground sloths, elk, giant chipmunks, and bison. Judging by the behavior of elk and bison elsewhere, the grazers followed fire, moving across the land to find the new growth that sprouts after a burn. Today, a few rabbits and voles nibble at the grass, and white-tailed deer sometimes wander out of the woods to pick at choice leaves. But grass here is largely bereft of its mammalian companions.

The loss in the southeastern US is part of a global pattern. As humans spread around the world in the last one hundred thousand years, grazing mammal populations crashed, sometimes to extinction. The extent of the die-off varied, from an 83 percent loss of grazing species in South America to a 22 percent loss in Africa, with other continents losing from 44 to 68 percent. These wild creatures have been replaced by agricultural grazers. Today, livestock outweigh all wild mammals on Earth combined by nearly thirty times. Intensely managed cropland has replaced natural grasslands in areas where the soil is fertile, while grasslands on less productive soils are, with a few exceptions, managed for livestock production on open range or steppes.

Lacking grazing mammals, about the only way for natural grasslands to persist is with the help of fires set by lightning and humans, especially in areas like Panola that have no domesticated grazers. Over much of North America, and especially in the southeast, Indigenous peoples used fire to regenerate grasslands and grassy open woods. This practice continued the long relationship between fire and grassland, this time mediated not by grazing teeth but by human ingenuity. When colonists arrived, they robbed and dispossessed Indigenous people of much of their land and put an end to traditional burning practices. In many places, dense shrublands and forests swiftly took over what had been open grasslands.

Today, many of the "natural" areas of the southeast are wooded, but this is a recent anomaly and distorts public perception about land management. Prairies belong in the American West, not at Panola Mountain or other places in the humid east, right? Not so. Some textbooks perpetuate this myth, claiming that grasslands grow only in places too dry for forests. But eastern North America never had a continuous forest canopy, especially not in the fire-prone South where grass diversity is high. Midwestern prairies have just two species of tall plumelike bluestem grasses. The South has nearly two dozen. At least one hundred million hectares of grassland and grassy open woodland grew across the eastern United States, despite the wet climate. Atlanta gets fifty inches of rain a year, more than Seattle. Paradoxically, this wetness does not dampen fire. Instead, rain feeds lush spring and summer growth of grasses, then the sky dries for a few weeks and the ancient love affair between fire and grass reignites.

"Have you heard of Burner Bob? The bobwhite quail?" Gabe asks me. "It's a new campaign meant to replace Smokey Bear [the US Forest Service's anti-fire mascot]." Forget the phoenix. After a fire, bobwhite quail rise from the ashes. These birds love bunchgrass like broomsedge, bluestems, and wiregrass. The gaps between the grass clumps make perfect quail runways. The bobwhite quail has become a pro-burning mascot of grassland restoration in the southeastern United States. Bob's mission is urgent. Since the 1960s, bobwhite quail have declined by about 3 percent every year, a population trend similar to that of other grassland birds.

"It can be really tough for people to realize what things might have looked like. Especially here in Atlanta, where trees grow just so quickly, it's easy to forget grasslands and think everything should be a forest. But, yeah, I think we're doing a good job, and I think we're starting to see some change."

Birds can be ambassadors for grasses, helping us to reimagine the landscape. As Gabe and I stroll over the ashy knoll, we spy a few field and song sparrows plucking morsels from the ground. Thirty robins work the field, too, along with an eastern phoebe and a bluebird sallying for insects. We can't see any today, but meadowlarks and several other sparrow species also live here. Except for the phoebe, these are all birds that live in or near grasslands.

"'Hope' is the thing with feathers . . . / Yet—never—in Extremity, / It asked a crumb—of me," Emily Dickinson wrote. But for grassland birds, hope does ask something of us: a lit match.

Or not a match, but a mixture of diesel and gasoline dripped onto the dry grass from the metal nozzle of a large "drip torch" canister. Nearly a year after Gabe and I walked the grasslands at Panola, I returned with a group organized by Birds Georgia to witness fiery renewal. Once the fire got going, I was shocked that knee- and chest-high grasses could produce flames that reached up at least four meters and smoke that soared to the clouds. The sound, too, was astonishing. From a couple of hundred meters away the burning grass sounded like heavy rain on a plastic tarp. At closer range, the higher notes of crackling and popping dominated, the result of water inside the grasses bursting into steam and grass fibers vaporizing into explosive gas. When the fire reached small pine trees that had sprouted among the grasses, the saplings detonated with a sound like hundreds of dry bones rattling in a wooden box, a clatter that lasted just a few seconds before the pine was entirely burned away.

Red-tailed hawks and American kestrels patrolled the perimeter of the burn, picking off escaping grasshoppers and panicked rodents. An American woodcock zoomed from the fire's edge right across our heads toward the safety of woodland, its flight feathers whistling as it passed. Field and song sparrows took agitated short flights, gradually edging away from the flames. Fire renews grassland, but not without short-term cost to the animals that live there. Most flee, a few get killed, and many have to find new homes until the grass swards rise high enough to offer cover and food.

The prescribed fire was organized by the Georgia Department of Natural Resources, with assistance from Birds Georgia. The goal was to renew the grassland, killing young trees and burning off dead grass so that ashes could fertilize the unshaded new growth of springtime grass. In one day, they burned about eighty acres, a task that involved fifteen people fully decked out with firefighting clothes, each with a heavy backpack of emergency protective gear. The team actively managed the edges of the fire to keep it contained and to steer it so that the whole grassland burned. Earlier in the spring, an arsonist did a solo job nearby. They torched the old sand trap that Gabe and I had walked across, along with one other small field. The arsonist had no ability to control the flames or the embers that floated into adjacent areas, and it was only through the work of local firefighters, and some good fortune, that the forest and houses downwind did not also incinerate. Grasses don't care about risk or legality. Broomsedge and little bluestem were already poking up in the arsonist's wake. Two weeks later, so, too, were grasses in the prescribed fire.

Blue-green grass leaves lancing through acres of charcoal black:

These are the colors of renewal, the banner of old revolutionaries working together, fire and grass, with a mammal helping.

❦ ❦ ❦ ❦ ❦ ❦

On one of the hottest days of a heat-record-breaking June, I return to the burned area on Panola Mountain with a dozen volunteers for Birds Georgia, a group made up of three kids with their mother, a flock of college students, and two plant-enthusiast retirees. We start early, but by the time the sun gets above the trees, it's already baking hot. We've come to sow fifty pounds of native grass seed, specifically *Sorghastrum nutans*, a species with plumelike golden flowering heads that often grow taller than head height. Its common name among Western botanists is "Indian grass," although it is only one of hundreds of North American grasses that benefited from burning and other management by the Indigenous peoples of the Americas.

At Panola, the weather is appropriate to the task of sowing this grass. *Sorghastrum nutans* is well adapted to heat. Deep-rooted, this grass persists when others wither. The plants' leaves are sheened in heat-reflective silver-blue. When the inevitable fire comes, they're not just ready, but eager.

With long-handled cultivators we scratch away duff and try to loosen the surface of the red soil. We then scatter a handful of prickly seed onto the seedbed, scuffing it in with our boots. It has not rained here in a month, so we bear down hard to get the metal tines of the cultivator to break the soil surface. Our backs and forearms protest. Sweat runs freely. Once again, mammals are toiling in the loyal service of grasses. An indigo bunting watches and sings from the adjacent woodland edge. No doubt some of the seeds we spread will feed him this afternoon.

As I work my way across the slope, I pick through some areas where broomsedge grows in thick abandon. The tentative tufts of March are now ankle and knee high. The burn did its job, making space and fertilizing. Belowground, broomsedge roots reach down twice as deep as the leaves extend upward, holding soil in place. Atlanta's South River, which runs just two hundred meters downslope, will taste less mud from this knoll when the rain finally comes. Dotted among the broomsedge are blackened brambles, camphorweed, and dogfennel, plants that the restoration team have zapped with squirts of herbicide. Although these species have their ecological roles—bramble feeds bees and birds, and camphorweed and dogfennel are favorites of small pollinators—they can grow so densely that they smother other species if not kept in check for the first years of prairie restoration.

Wildflowers sprout between the broomsedge bunches. Mountain mints with flowers that look dusted with icing sugar grow waist high. I gently squeeze their leaves and smell spearmint sharpened with a splash of gin. Even in drought, mountain mint is a font of nectar. Bees, wasps, and hoverflies scramble over their clusters of blooms. Also gleaming among the silvery blue-green of the grasses are flowers of lemon yellow and purple-centered gold, sneezeweed and plains coreopsis. Skipper butterflies dangle from the blooms. Two black-eyed Susan plants are in flower and I spot the pealike leaves of several wild indigo plants. All these plants were seeded or transplanted by Birds Georgia, with help, in the case of mountain mint, of seeds that blew in and were dormant in soil. In adjacent meadows, restoration is more advanced and flowers of rosinweed, dwarf sumac, milkweeds, bee balm, and wild indigo plants are abundant.

I'm delighted to see so many wildflower species here, but the species count of plants is low, about 10 percent of what we'd encounter in an old-growth grassland, one that had not been plowed or forested in centuries. It takes decades or longer for degraded grassland to regain its diversity. Yet, in just three years since this restoration started, a former golf course and construction site has turned into a young prairie, its diversity rising each year.

The emergence of wildflowers and herbs here reveals one of the grasses' superpowers. They build community. The term "grassland" is misleading. Although about 5 percent of the eleven thousand grass species worldwide naturally grow in near monocultures—saltmarsh grasses are one example—most grasses create conditions in which other plants thrive. By keeping woody competitors away, grasses allow other plants to bask.

Grasses are community builders. They often host more plant species than any other habitat. The world record for the greatest number of plants crowded into one square meter is a meadow in the mountains of Argentina, with eighty-nine species per square meter. In the grasslands of the southeastern US, on the scale of hectares, the richness of plant life rivals that of the Amazon rainforest, although at larger scales the tropical forest is more diverse.

Unless you're kneeling and peering closely, it is hard to see grassland diversity. An old-growth grassland has less power to awe human senses than big trees in an ancient woodland. Occasionally, we get a glimpse. Some of the most spectacular natural displays of wildflowers are in grasslands: Texas meadows filled with the five species of lupine known as bluebonnets. "Superblooms" of poppies and dozens of other species in the California hills. Hay meadows and grazed pastures in England filled with wildflowers, including the rare species that kindled my mother's love of orchids in the Chiltern Hills. Across the UK and much of Europe, these meadows are the only home of many flower species. Rainy season floral extravaganzas in the grasslands and savannas of Tanzania, Kenya, and Uganda. Mediterranean grasslands and grassy woods filling with wildflowers in late spring. At higher elevation, in the Alps, Western Ghats, and Rocky Mountains, montane grasslands host profuse summer blooms of wildflowers. Lower-elevation prairies and meadows grow on every continent except Antarctica, and are home to thousands of plant species that live nowhere else, including the goatsbeard and orchids we encountered earlier in Washington state and France. Want to see thousands of lovely flowers? Head to a grassland.

Plants in grassland coexist because they specialize. Virginia wild rye grows in the cooler parts of the year; big bluestem thrives in the summer heat. Reed canary grass and eastern gamagrass do best in wet areas, but little bluestem and broomsedge prefer dry. Black-eyed Susan grows and blooms earlier in the year than goldenrods and most asters. Roots divide up the underworld. Some, like blazing star and indigo, snake their roots into the soil two meters or more deep. The roots of most grasses grow in a dense fibrous layer near the surface, with sparser tangles descending more than a meter. Plants also differ in their feeding styles. Legumes like clovers and indigo nurture specialized bacteria in root nodules that grab nitrogen from the air. This nitrogen eventually circulates through the soil and flows to other plants. Most plants slow or shut down their growth in summer's heat, but some bunchgrasses persist even in drought. Grasslands may look uniform, but plants have made thousands of nooks and crannies. A grassland is a multidimensional universe: To the three dimensions of space, plants add rhythms in time, the geometries of growth forms, and the intricacies of how they feed on sunlight and soil.

Out of diversity comes productivity. When scientists artificially grow grassland communities with different numbers of species, they find that the most diverse communities are the most productive. When conditions get tough, the advantage is magnified. Like investing in the stock market, a broad portfolio is best. No wonder some mammals evolved into grazers. Grasslands not only provide abundant forage, they do so consistently. Varied plant life also gives grazers choices. Depending on their needs, they can nibble nitrogen-rich legumes, get a grassy carbohydrate hit, then cap it off with some tannic tick trefoil to help digestion, or medicinal coneflower and leadplant. Productivity supports insect life, too. As I walk through the broomsedge in June, dozens of grasshoppers clatter away with every footstep. Bees and wasps wing past, leafhoppers spring, and beetles scurry for cover. This productivity is why so many birds depend on grasslands for their breeding or wintering. Grasslands, especially those in humid areas with good soil, provision their local food webs as richly as do forests.

Grasses also build soil. Their leaves send about two thirds of all the food they make to the underworld. There, roots tunnel many meters down. As they grow, they break up clay and rock, exude sugars and other molecules, and interweave their cells with fungi. When the roots die, they add spongy organic matter to the soil. This soil-building process is so productive that it lifts the ground. When a degraded grassland returns to health, the ground heaves up, as if inhaling with relief. In old grasslands, the soil can be rich with organic matter to a depth of several meters. When prairie goes under the plow, most of the organic matter disappears, turning living water-holding, nutrient-rich soil into mineral dust. Today, despite widespread degradation of grasslands, one third of all carbon stored on land is still locked up in grassland soils.

As fellow volunteers, the staff of Birds Georgia and I sow grass seed, we enact the grassland ethos: Build community, one species helping another. Grasses are creators. They use cooperative partnerships to build their homes, places that, in turn, open possibilities for others. They hoard soil carbon, create habitat for other plants, and feed animals.

Although we don't often imagine ourselves in this way, we are a prime beneficiary, a species built by grass.

❧ ❧ ❧ ❧ ❧ ❧

What's for dinner? Grass. Wherever you live, some kind of grass is probably feeding you.

When the prophet Isaiah proclaimed that "all flesh is grass," he intended a commentary on the fleeting nature of human life, but he also spoke an ecological truth. In Isaiah's time and in ours, grass sustains us. If we stacked in 50-kilogram sacks the total cereal harvest in 2023, the pile would reach to the moon forty times. That's 2,836 million metric tons of grass flowers matured into seed. Three grasses—rice, maize, and wheat—account for 90 percent of this superabundance, supplying us with two thirds of food calories. The juices of sugarcane, another grass, supply another 1,900 million metric tons. Barley, sorghum, oats, millet, rye, and wild rice are grasses, too. Livestock fattens on grass from pasture and the maize-filled troughs of feedlots. While our great ape cousins feed on forest fruits, leaves, and animal prey, we depend on grasses. If we named ourselves for our primary food, we would be grass apes, *Homo poaceae*, for Poaceae, the scientific name for the grass family, from the ancient Greek for "fodder."

It is the nutritive gifts of grasses, with help from oil-rich fruits like mustards and oil palms, that caused the increase in food calories available to humans over the last millennium and, especially, the last century. The cereal harvest in 2023 was 50 percent higher than that of 2000 and three times that of the 1960s, outpacing human population growth on all continents. Famines are rarer than they were and now largely emerge from human injustice and war, not the failure of plants to yield food. Such productivity comes with severe costs: felled forest, mined and synthesized fertilizer, among many. But those who in the nineteenth and twentieth centuries made erroneous predictions of imminent mass starvation erred by underestimating the world-changing potential of grasses.

These global patterns are evident in kitchens. At home, the bottom drawer of our kitchen cabinet grinds when I pull it open. The poor thing has worn sliders and is loaded with bags and tubs. Bread flour, whole wheat flour, all-purpose white flour, masa, purple cornmeal, medium-ground yellow cornmeal, plain fine cornmeal, semolina flour, barley flour, and sorghum flour. Some are baking staples, ingredients for pancakes, loaves, and corn breads. Others are aspirational, plucked in moments of enthusiasm as Katie and I push our cart through the aisles of the Dekalb Farmers Market, a bustling warehouse near our home stocked with bulk dried goods and fresh produce from across the globe. Regardless of their origin, every one of the flours in our kitchen drawer is ground-up grass seed, the product of a mature grass flower. Other kitchen drawers hold rice and pasta, also made from grass seed. Our kitchen, like kitchens over much of the world, is a bouquet of grass.

From the three hundred thousand species of flowering plants on Earth, we've plucked a handful of grasses and founded modern agriculture on their productivity. What made grass so special? The answers reveal not only why we latched onto them so firmly, but also how grasses managed to take over much of the planet long before humans evolved.

Grass flowers are super-mothers, giving their embryos ample provisions. Under a magnifying glass we can see how. I pull open the complaining kitchen drawer and dip a teaspoon into some bags, retrieving flours that I dust onto scrap paper under a bright counter light. What looked to my unaided eye like powders of different colors reveal themselves under the lens as diverse and beautiful. I expected white flour to look fluffy, but magnified it looks like coral sand. I smooth the tiny pile with the back of my spoon and the flour becomes a miniature tropical beach, a gleaming expanse enlivened with a scattering of darker grains. Whole wheat flour seems made of tan-colored sand mixed with shredded cardboard, as if a hurricane had passed through a shipping warehouse on its way to the beach. The grains of purple cornmeal are larger than those of the wheat flours and are intermixed with white-blue pebbles and chunks of broken obsidian. Uncooked rice grains loom over these sands. They are slightly translucent and etched with lines, as if ancient Egyptians had built their obelisks from milky glass. Who needs magic mushrooms when we have 7× hand lenses?

Most of what I'm admiring under the lens are not the tissues of the seeds' embryos, but the ground-up remains of the lunch box that the mother flower carefully packaged and gave to each of her departing children. White, tan, and purple "sand" and the glassy bulk of the rice grain are all what botanists call "endosperm," a tissue that feeds the embryo when it germinates. Endosperm is only found in flowering plants, although a few close relatives like the nonflowering shrubby joint fir, *Ephedra*, have its rudiments. The evolution of endosperm was one of the innovations that gave early flowering plants an edge.

Orchids and a few other flowering plants secondarily lost or reduced their endosperm when they evolved tiny seeds. Without endosperm, orchid seedlings are entirely dependent on fungi. Grasses went the opposite way, swelling the endosperm to such an extreme that the embryo looks like an afterthought, a bundle squashed into one end of the seed. Such ample provisions give grass seeds a measure of independence, at least at first. If you've ever sprouted wheat grains, you know that grass seeds can lance roots and baby leaves with no help from fungi or other partners. Later in life, community building becomes more important for grasses, but their seeds can go solo.

It takes strange sex and disturbing sibling relations to make endosperm. When pollen grains hatch on the female stigma of a flower, each pollen grain first grows a tube toward the egg, then sends a pair of tailless sperm down the tunnel. If the female part of the plant decides to accept them, both merge with the female tissues. One sperm unites with the egg, making an embryo. The other unites with ladies-in-waiting cells nestled alongside the egg. This second fertilization forms a sibling that never develops root or stem, but instead fattens with starch and other food, all drawn from the mother flower. This is the endosperm. Every seed of a flowering plant carries two genetic individuals. One, the embryo, can grow into a new plant. The other, the endosperm, is an undifferentiated mass of cells whose fate is to feed its sibling and die in the process.

I look on my piles of flour with horror. Not only am I admiring pulverized embryos, most of what I'm seeing are the remains of creatures that, thanks to evolution, grew into starchy blobs destined to be devoured by their siblings. The mushroom trip just took a bad turn. At lunchtime, hunger overrides my qualms. Endosperm makes excellent bread, rice, or tamales. Thank you, doomed siblings.

Among agrarian humans, endosperm left its mark on our genomes. People whose ancestors ate a lot of endosperm have extra copies of the genes that make starch-digesting enzymes. Evolution has built dependence on wheat, corn, and rice into some human DNA. Grass lives inside us. It also linked us into new mutually beneficial relationships with other creatures, as flowering plants so often do. Fermentation by bacteria and fungi is central to many grass-based human foods, including wheat bread, beer, and dozens of fermented rice and corn dishes. Grass, microbes, and humans are a powerful trio of cooperators.

Endosperm reserves made grasses especially good at thriving in challenging conditions. Once tucked inside the seed coat, the starches, oils, and protein in endosperm keep for months. Endosperm is the original dry pantry. Grass seeds can bide their time, waiting for the right conditions to germinate. When rains and warmth arrive, the endosperm digests itself and shunts food to the growing seedling. Powered from within, grass seedlings lance out roots and stems, beating any competition. It is no accident that many of the most successful recent transcontinental transplants of modern plants are grasses, often causing problems in natural habitats and agricultural fields as they push out local species or desirable crops. Looking further back in time, the fossils and family trees of grasses show that they've been colonizing new lands for millions of years, largely thanks to the endosperm in their seeds. These Vikings are successful not because they are warlike, but because their mother flowers sent them out with food hampers.

Like a duckling that emerges from the egg ready to feed itself and scurry away from predators, grasses are ready to take on the world from the moment they "hatch." Unlike the mostly undifferentiated embryos of their close relatives, grass embryos have roots, miniature stems and leaves, and a placenta-like structure connecting to the endosperm. Once conditions are right, the embryos are ready to burst into action. A rice grain or wheat kernel in your hand might look passive, but you're holding a time capsule with a sprinter coiled inside.

What I've called "seed" or "grain" is technically the fruit of the grass flower. Like all other flowering plants, grasses wrap the embryo and endosperm in layers of maternal tissue. Unlike fleshy fruits like plums or mangoes, the fruit tissues of grasses are thin and papery. In whole wheat flour, these fibrous coverings are called the "bran," the cardboard-like material that I saw under my lens. The "wheat germ" is the embryo, little flecks of chewy brown. White flour is nearly pure endosperm. Like drinking cow's milk, when we eat white bread or pasta we're getting a slug of energy and nutrients built by evolution for the rapid growth of infants. Perhaps pasta or pizza with cheese is so comforting because we're literally being mothered by flowering plants and bovines?

Not content with a tightly bundled fruit, grasses add extra papery layers around the seed, like gift wrap gone mad. A swarm of botanical terms describe the details of these wrappers—floret, glume, lemma, awn, palea, lodicule—but we usually refer to them as mere chaff, the fibrous junk that stands in the way of our hunger for endosperm. This name belies the intricacy of each species' arrangement of parts. Some look like nested canoes, others like sheathed javelins. Many are arranged in geometric patterns of fans or rows. Each arrangement serves to launch the young on a journey suited to the ecology of its species. Many have wind-catching bristles. Some are grappling hooks for mammal fur or bird wings. I once made the mistake of wearing wool socks to walk through grassland at Panola Mountain and later spent hours unpicking hundreds of sharp, grippy grass seeds, a reminder of how well they cling to fur. Other grass seeds twist as humidity changes, drilling seeds into the soil. The collective term for all these fruit wrappers built by the mother flower is "spikelet," a feature that evolved in the ancestor of modern grasses that has literally carried them, by wind and animal bodies, to worldwide success.

These winning characteristics of grass seeds were lately helped by a few quirks that made them especially alluring to humans. We have delicate guts, unable to tolerate the poisons and tannins that lace the seeds of many other flowering plants. For example, a few seeds of larkspur would kill you by alkaloid poisoning. Without repeated soaking in water, acorns sicken us with bitter tannins. Our saliva has only weak levels of the tannin-neutralizing chemicals that other mammals use to tolerate bitter food. Compared with the seeds of almost every other plant, grass seed is highly palatable, especially for a dexterous ape whose hands can slough off the chaff. Many grasses evolved to be partly dispersed by herbivorous mammals, the seeds passing through guts with the bulky leafy tissue. For millions of years, then, it has been in the interest of grasses not to have poisonous seeds, although many have tough or bitter coats to deter fungi, insects, or overenthusiastic grazers. When our ancestors started to gather and replant seeds, grasses were the obvious choice: edible, portable, rot resistant, and stuffed with easy-to-digest food.

When grass fruits mature, they usually detach from the parent, hitching a ride on an animal or the wind. Botanists call this process "shattering," because a single touch of a ripe seed stalk can cause all the fruits to snap and tumble, like shards of broken glass. The mother plant encourages this fracture by killing and partly digesting the cells where the mature fruits attach to her stalk. In some grass species shattering can be turned off by just one or two gene mutations. The mother fails to cut the umbilicus and the fruits cling tight. In the wild, these mutants are hopeless, failing to send offspring into the world. But when clever bipeds evolved, our ancestors, the doomed mutants became, literally, the seeds of the agricultural revolution. By not falling to the ground, the shatterless mutants stayed on the stalk until we were ready to harvest them. By selectively keeping and replanting the mutants, ancient humans gained the first cereal crops. Today, when we bite into a wheat sandwich, a corn tamale, or a rice curry, we're eating descendants of these shatterless mutants.

Other mutations, random genetic changes that humans noticed and kept for planting the next year, helped along the way, notably those that multiplied the number of fruits on the stalk and fattened each seed. Many cereal crops also went through repeated rounds of genome duplication and hybridization, just as goatsbeard did. But without early shatterless mutations, our species would never have enmeshed its fate so deeply with grasses. Imagine a world where rice, wheat, or corn remained minor, "wild" foods. Agriculture would be vastly less productive. Human population size would be a fraction of what it is today. Few genetic changes have been as consequential for the story of life on Earth.

The shatterless mutations launched humans on our present trajectory. We've been clearing forest and tilling fields for grasses ever since. The book of Genesis, the origin story of a grass-based agrarian culture, tells us, "In the sweat of thy face shalt thou eat bread." God decrees: Humans toil on behalf of grass. But our dependence on grass far predates the origins of agriculture or written scriptures. Without grasses, we'd still be small-brained apes living in the trees.

❦ ❦ ❦ ❦ ❦ ❦

To get a clearer view of the early days of human evolution, I'm doing a kitchen experiment. Wheat and corn seedlings sprout from a thin layer of soil in jars on our windowsill. I sowed them a week ago and already the little Vikings' leaf tips have reached the top of the jar, a vivid demonstration of how endosperm gets mobilized for rapid growth. A young orchid, with no food supplies from mother, would take a couple of years to get so large.

Wheat and corn belong to two different clans of grasses, groups that split from one another about sixty-five million years ago. Members of the wheat clan are mostly adapted to growing in cool, wet weather. The corn clan generally thrives in heat and, often, drought. Although the seedlings look superficially similar, they have a hidden difference, one that plays an out-sized role in the evolution of our species. Human evolution, it turns out, partly hinges on grass physiology, a strange and consequential convergence of the storylines of plant and primate.

I reach into the jars and pluck a couple of unfortunates. Using a box-cutter blade, I slice each one with an oblique cut and put it on a microscope slide. I feel like a character in a Jules Verne story, using pre-digital technology to dive into wondrous unknown depths. Magnification is submersion. Through the eyepiece, I see a sliced leaf the way I might gaze at normal scale at a cliff face. The effect is helped by the equipment I'm using, an early twentieth-century brass-and-steel pocket micro-scope with a steampunk vibe, one of the few material possessions my fa-ther kept from his youth and that I inherited. One hundred years after it was made, the microscope still delivers a good image, albeit one pep-pered with imperfections from dust and a flaked mirror.

My first impression of the magnified cut leaves is that they are on fire with green. Illuminated from below, thousands of blazing emerald globes are crammed across each leaf. These are the chloroplasts, sacs stuffed with chlorophyll. No wonder the seedlings grow so fast: Light has no chance of passing through without tangling with chlorophyll pigment.

My second impression is how orderly the insides of the leaves are. Like bricks on a walkway, cells line up end to end, running lengthwise, parallel to the leaf edges. The seedling leaves are barely as wide as a strand of spaghetti, yet each has nearly one hundred parallel lines of cells running from leaf base to tip.

The rows of cells are especially evident at the ragged sliced edge. There, the cells are like clear glass, lacking all pigment. This is the leaf's transparent skin. Mixed among these surface cells are glassy spikes, hairs that protect the leaf from insects. On the lower surface of the leaf, the skin is studded with mouthlike structures, slightly larger on corn than wheat. These are breathing pores, made from two cells that act as lips. When they part, air and water vapor flows to the interior. Grass lips are especially energetic compared with those of most other plants and can open and close in minutes depending on whether the plant needs to inhale carbon dioxide or seal up to save water.

Pipes run along the length of the leaf, parallel to the brickwork. These are conduits for water and food. Each pipe is sheathed with a layer of cells, like the foam insulation I slip around exposed pipes in our basement. In wheat, these wrappers are translucent and relatively small. But in corn, the wrapper cells are fat and packed with green globes, chloroplasts. What seems like a minor difference in plumbing wrappers is in fact a revolutionary change. Corn's chlorophyll-filled sheath cells allow it to thrive where other plants fail. In the past, this new form of plumbing remade many ecosystems, including the African savannas in which our ancestors evolved. In the future, these special wrapping cells will increasingly feed us in a hotter, more drought-prone world.

These are grandiose claims for a pipe wrapper in an ordinary blade of grass. To understand why these cells are so important, let me introduce one of the most important chemicals on the planet, one most of us have never heard of: rubisco. The sheath cells in corn and its kin serve this molecule.

Rubisco—a mercifully shortened version of a longer chemical name—is the protein that grabs carbon dioxide from the air so that plants, bacteria, and other photosynthetic creatures can knit carbon dioxide into food molecules. Almost every bite of food we eat depends on rubisco. It's likely the most abundant protein enzyme in the world, about seven hundred million metric tons of it divided among billions of cells.

Up to 10 percent of the dry weight of the seedling leaves under my microscope is rubisco.

All hail rubisco, the foundation of earthly life. A venerable old molecule, feeding ecosystems since it evolved at least 3.4 billion years ago. Also, a horribly inefficient molecule. Rubisco is sluggish, cranking at a mere 1 percent of the speed of other enzymes. Worse, it makes mistakes. Lots of mistakes. In a happy leaf growing in moist, cool conditions, rubisco misfires about one third of the time. It grabs oxygen instead of carbon dioxide. This squanders time and makes toxic and expensive-to-recycle waste. In a hot, water-stressed leaf, rubisco does nothing but misfire. The molecule on which most of life depends is a clunker.

When rubisco first evolved, misfiring wasn't a problem. On the primordial Earth, carbon dioxide was twenty or more times more abundant in the atmosphere than it is now, keeping the Earth greenhouse toasty. Soaked in superabundant carbon dioxide, rubisco seldom erred. More recently, when flowering plants appeared and diversified, carbon dioxide was up to four times more abundant than it is now. Rubisco mostly chugged along happily. It is only in the last fifty million years, and especially the last twenty million, that carbon dioxide levels have sagged. Like an old engine unable to run on modern fuels, rubisco shudders and sputters, and sometimes grinds to an inglorious halt. Look around you on a summer's afternoon in a city park. Many of the plants have stopped or drastically slowed their feeding on light. They've closed their breathing pores to conserve water and their rubisco has seized up, starved of the meager supply of carbon dioxide that kept its cranky mechanism running in the cooler morning hours.

But not all plants in the park have shut down. If the park has a community garden growing corn or a meadow of grasses native to warm climates, they're likely still making food in the afternoon heat. Inside the leaves of corn and its kin, cells in the main part of the leaf grab carbon dioxide and, instead of using it themselves as they do in wheat, attach it to a carrier molecule, then ship it to the sheath cells around the veins. There, the carrier drops its load, creating a rich soup of carbon dioxide. This pump re-creates the primordial earth around rubisco in the sheath cells. Bathed in superabundant carbon dioxide, rubisco purrs along with few mistakes. By concentrating carbon dioxide around rubisco, the corn plant pumps out food when other species falter. Biologists call this turbocharging process C4 photosynthesis because the carrier is made from four carbon atoms.

C4 turbocharging is such a great idea that it evolved at least nineteen times in flowering plants. Grasses are the most successful and consequential of these innovators. Today, although C4 plants represent fewer than 2 percent of plant species, they do 25 percent of all photosynthesis on land.

The slivers of leaf under my microscope are representatives of two great clans of grasses, thousands of species in each. Corn is a C4 plant and, like all its close kin, thrives in hot and dry conditions where rubisco is especially error prone. Broomsedge and bluestems also belong to this clan, as do most grasses in the subtropics and tropics. Wheat belongs to the other clan, grasses that rely on unaided rubisco, so-called C3 plants named for a three-carbon molecule in their chemistry. C3 grasses thrive in the cooler, wetter regions. Turbocharging is great when you need high performance in the heat, but the old C3 pathway still works well when plants grow in temperate conditions. Many C3 grasses also have antifreezes and other winter defenses that most C4 plants lack. When wheat and other C3 plants are grown in hot areas, they are usually planted as winter or spring crops, doing most of their growth during the cooler, wetter parts of the year. C4 crops, on the other hand, are heat lovers, as anyone who has grown corn in their garden can attest. During the hottest days of summer, when other plants are struggling, corn surges.

Between them, the two grass clans are now the calorific foundation for human life worldwide. Wheat and rice, both C3, feed us in the temperate world. C4 crops like corn, sugarcane, sorghum, pearl millet, and tropical pasture grasses give us food in hotter places and will play an increasingly important role as the planet warms. Geneticists are cajoling rice and other C3 crops to become C4, so far with limited success. These attempts carry strong echoes from the earliest days of human evolution. By turning to C4 plants for food in a rapidly changing world, we are re-enacting an ancestral dependence. C4 grasslands in Africa made humans. Without them, we'd still be arboreal creatures, foraging for fruits in the forest, or small-brained savanna apes, poking around for tubers and termites.

Tooth enamel tells the story. The balance of C4 and C3 plants in an animal's diet leaves a signature in the carbon atoms of the body. By a quirk of their physiology, one with no functional significance to the plant, C4 plants selectively accumulate slightly heavier carbon atoms, those with an extra neutron. This means that we can infer an animal's diet from the balance of carbon atoms in its body. For carnivores, the atomic signature reveals what kind of vegetation their herbivorous prey ate. A corn-eating vegan and a carnivore subsisting on corn-fed beef both have carbon atoms that say "corn." "All flesh is grass" is true down to the level of carbon atoms. For long-dead animals, tooth enamel preserves the atomic signatures of the animals' diets. Much of what we know about the diets of animals in the past is inferred from carbon in teeth.

Tooth-snared carbon reveals that, over the last ten million or so years, some mammals on every habitable continent moved out of the forest into expanding C4 grasslands. The newly evolved mammals accelerated the transformation of forests into grasslands and grassy savannas, with help from grasses' old companion, fire. The timing differs among continents, with C4 grasses becoming common whenever heat and highly seasonal rainfall promoted fire. On one continent, Africa, the changes were especially pronounced. Starting about ten million years ago, horses and zebras swiftly became C4 specialists. Others, like hogs, hippos, and giraffes, took longer and fed on a mixture of C3 and C4 plants.

Among the animals to move into grasslands in Africa were the great apes, including some of our ancestors. Compared with most large grazing mammals, these African apes were rare. They'd barely be worth mentioning if not for the later ascendancy of one of their number. Starting about five million years ago, these creatures knuckled on all fours and, later, loped on two legs, into the grassland. Just as there are many species of antelope and other herbivores, so, too, with hominins, our ancestors and their extinct relatives. The hominin family tree back then was a bush, not a single line leading inexorably toward *Homo sapiens*. Multiple species coexisted, each an evolutionary experiment in how to turn a tree-swinging ape into a walking grassland humanoid.

As with large herbivores, the hominin ape species differed in their embrace of C4 grasses. *Ardipithecus ramidus*, a species that lived 4.4 million years ago, had a C3-rich diet similar to that of chimpanzees. *Paranthropus boisei*, an especially stout-jawed species that lived from 2.5 million to 1.15 million years ago, fed almost exclusively on C4 vegetation like grasses and sedges, or on the animals that fed on these grasses and therefore had C4-rich flesh. Others, like *Australopithecus africanus* (3.3–2.1 million years ago) and *Paranthropus robustus* (2.3–0.9 million years ago) had mixed diets and, in *africanus*, a great deal of variability among individuals. Although we cannot know for sure, this variability suggests human-like traits such as individual and cultural preferences for different foods. Did some parent *Australopithecus africanus* despair at their kids' refusal to bite grasshoppers or chew grass seed? Did they look with fascination and sometimes disgust at the weird food preferences of those "other" hominins, the ones living on the lakeshore or deep in the grasslands?

The oldest physical evidence of our genus, *Homo*, is a 2.8-million-year-old fragment of a jawbone from Ethiopia. Blackened teeth, still shiny after long burial, jut from eroded bone. Inside the teeth, carbon atoms retain the signature of both C3 and C4 grasses, an imprint very much like that of modern savanna grassland animals such as giraffes and some antelopes. Most likely this prehuman ate plants and animals from forest edges, and from C3 and C4 grasslands. Later members of the genus *Homo*, presumably among them some of our direct ancestors, continued this middle path. Then, from about 2 million to 1.4 million years ago, their diet shifted, or so it seems from hominin remains from the Rift Valley of Kenya. Their tooth enamel shouts: Bring on the C4!

The sudden embrace of C4 plants by these prehumans was a choice. The plant community in which they lived had not much changed. C4 grasses were already well established in the region. Rather, these ancestors somehow decided to focus their attentions on the plants and animals of the C4 grasslands. They may have eaten grass seed, maybe even some grass leaves, and a few other C4 plants like sedges from wetlands. Perhaps C4-eating insects were part of the mix. Judging from butchered remains, these ancient people also figured out how to team up and successfully kill dangerous wildebeest-like C4-eating bovids. The shift to a C4-rich diet coincided with the development of the first oval, bi-faced stone hand axes and cleavers, technologies that replaced cruder flaked stone tools. This was a momentous change. No other primates regularly hunt cooperatively to take down large prey and none use sophisticated tools to do so.

As technologies advanced, evolution reshaped the hominin body, enlarging the brain, flattening the face, and shrinking chewing teeth and jaw muscles. The species that emerged during this transition, *Homo erectus*, is our direct ancestor and the first hominin to make extensive migrations out of Africa to Europe and Asia. *Homo erectus* was also the first hominin to control fire for cooking. From then on, the hearth was the center and catalyst of human culture. Grasses, mammals, and fire: The old partnership was, yet again, changing the world.

The sliced grass leaves under my microscope are marvels not only for their beauty and for what they reveal about the inner workings of leaves. They also disclose part of our creation story. This narrative is not a solitary hero's journey, humans rising alone, driven by the force of our will and ingenuity. Rather, the fruitful union of grass and ape storylines made us, helped by fire. We owe our bipedalism, tools and tool-inventing brains, and socially sophisticated foraging and hunting to grasslands, especially those with C4 grasses. Our almost total dependence on grasses for food today is a continuation of a much older tale, as will be our future need for heat-tolerant C4 crops.

Cartoons of human evolution often show apes gradually unbending and striding into the future, the so-called March of Progress. The images err because they imply a singular, linear path rather than a many-branched genealogy. The cartoon should also be thick with grasses. We did not stand up alone. Grasses lifted us. Truly, we are grass apes, *Homo poaceae.*

Entanglement with grasses on land made us and will continue to sustain us long into the future. Lately, our fate, and that of many other living beings, also hinges on another grasslike plant, one that grows mostly unknown and uncelebrated in the muddy margins of our terrestrial world. Long before the rise of grasses on land, seagrasses, an entirely different clan of flowering plants, transformed the seas. To this day, they sustain ocean and terrestrial life worldwide.

Seagrass

Muddy saviors

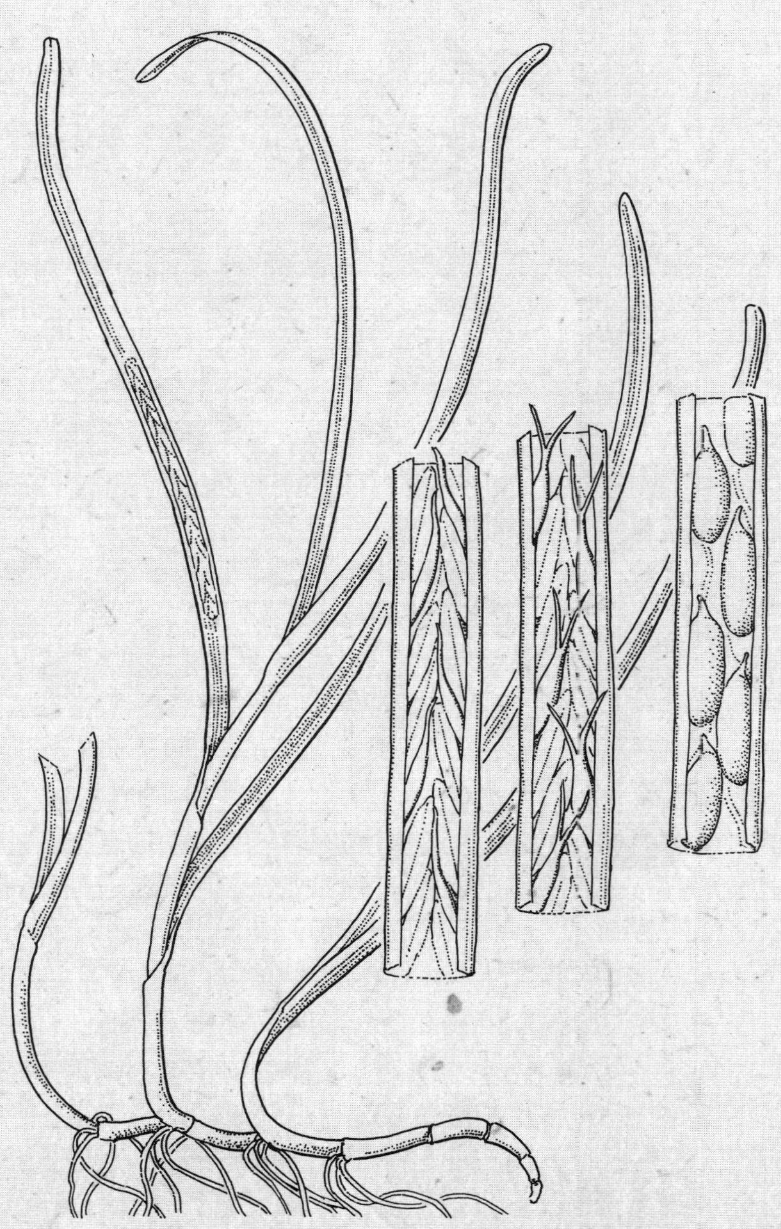

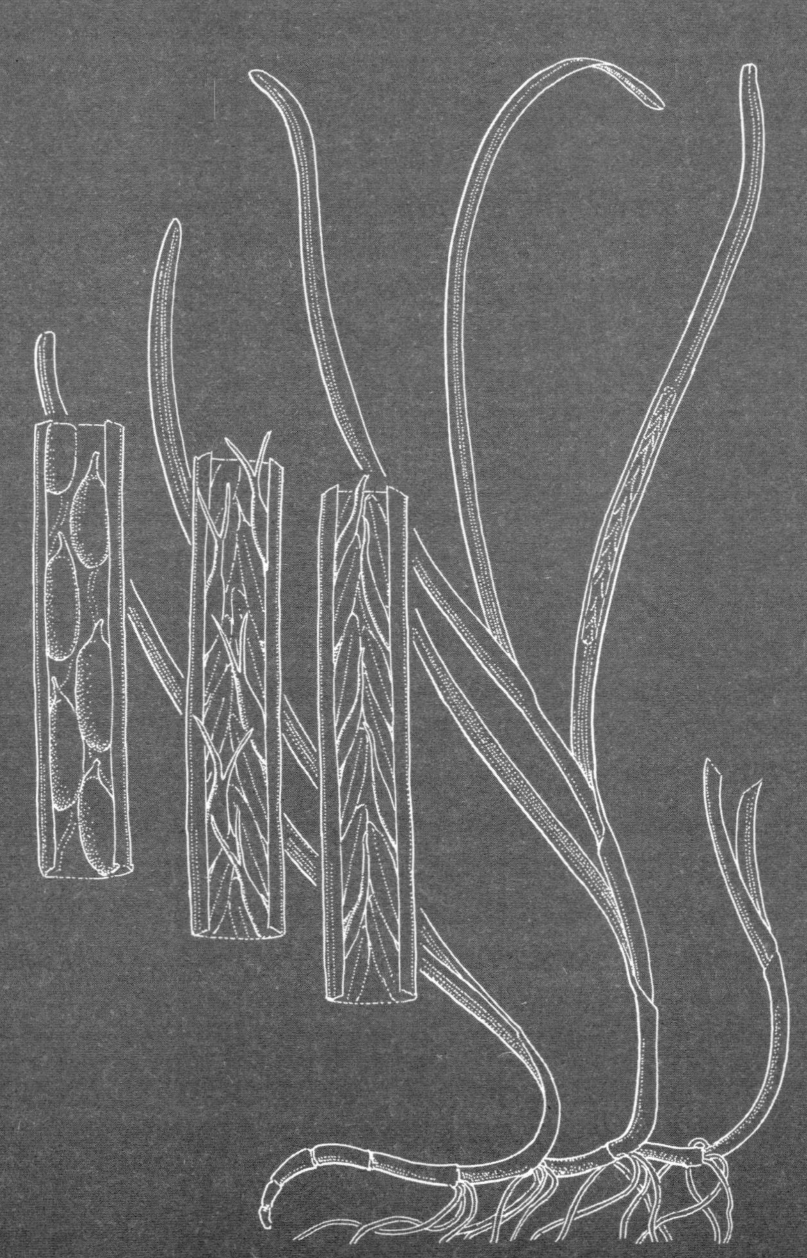

BLACKNESS CASTLE LOOKS LIKE A STONE WARSHIP headed to sea. Its walls form a bow and stern, with mast-like towers and a raised quarterdeck. Built on a rocky spit that juts into the Firth of Forth estuary in Scotland, the castle served for five hundred years as a garrison, prison, munitions store, and royal fortress.

My sister, Julia, and I park down the road from the castle, near the pub. She lives at the foot of the Pentland Hills, inland, and has cheerfully indulged my wish to drive out to the Forth, even though it's going to be "bloody freezing out there" in a wind gusting straight down from the Highlands. We have come on a brisk spring morning not for military history, but for plants creeping through mud. Because of the castle's long-standing importance, the spit has been reinforced, creating small bays on either side, each a little under a kilometer across. Over the centuries, gunk has settled on the bottom of these shallow tidal waters. At low tide, the warship seems lodged in mudflats. In this dark ooze grows seagrass, a flowering plant that was, until recently, one of the most obscure and unloved plants on the planet. Unknown, that is, except to members of a few Indigenous communities who have long understood that seagrass is a world-maker, knowledge that Western scientists and conservationists are finally rediscovering.

Julia and I reenact delights from our childhood: wellies on, warm hats on, stomp out into the muck. "Here's some, I think," she calls, grinning. I lope over in mud-weighed boots. Slender green ribbons stretch across the mud's surface. When the tide rises, the buoyant ribbons will rise and wave with the currents. A few are as long as my forearm, but most are a handspan tall. At their base, they are pearly lime green, darkening to pine-needle green along most of their length. When I lift a strand, it is slightly translucent. This is *Zostera marina*, a type of seagrass known as common eelgrass.

"Are you sure that's not an alga? They don't look like much." She's right. The strands look like seaweed. Boring seaweed, at that—no frilly edges or air bladders, just pencil-thin strands of mud-splattered green. We see no flowers at all. Later, I learn that these algal-like strands have tiny flowers tucked inside folds, but to the unaided eye it is hard to believe that this is a flowering plant.

I crouch down and worm a finger into the mud. "Look, they go way down. I can feel roots and . . . ugh, that's tough! I can't budge it." Two knuckles down I find what feels like knotted sisal twine, rough textured and strong. The green ribbons sprout from this buried network which is also tough, resisting my tugging like leather straps. No seaweed grows like this. Seaweeds either cling to rocks with holdfasts or attach loosely to the surface of mud. Seagrasses are only very distantly related to grasses on land, but they've independently evolved similar body forms: short-lived photosynthetic blades growing from tenacious buried stems and roots.

"Once you get your eye in, you can see them from a distance. Look over there." Julia points across the mudflats, then swings her arm, counting off two or three spots. "I'll bet those patches of slick green stuff are seagrasses, not algae." Another childhood echo: I'm trying to pull my feet out of the mud while she's already figuring out how the community works.

The mud is especially deep and wet along the rivulets that drain the flats, so we pick our way around, taking a meandering path as we head to new patches of exposed eelgrass. Our wellies soon teach us that mud is firm wherever the plants grow. Viscous ooze encloses our feet when we step beyond the eelgrass.

Blackness village today is a quiet spot with a small boating club, a pub featuring seafood and a poetry contest, and a few dozen homes, mostly of retirees. Not much action comes off the Forth, save for the lively wind. But from the fourteenth to nineteenth centuries this was a bustling mercantile harbor, with a customshouse and warehouses. Salt, brick, tile, mined minerals, and seafood all moved through the harbor, which also served as the port for Linlithgow royal palace, just six kilometers inland. During times of conflict, warships crowded the harbor and the castle was an embattled stronghold, especially during the struggle between Mary, Queen of Scots, and Henry VIII, and later bombardment by Cromwell's army. By the mid-nineteenth century the harbor was in disuse and its piers crumbled away. Today, a small renovated pier serves the boat club, but sediment has reclaimed the rest of the harbor. What was described in 1710 by the physician and antiquary Robert Sibbald as "a harbour for all Ships" is now so shallow that, by picking the right path, Julia and I have walked at low tide through the center of the bay all the way to the level of the castle.

Julia's background is in engineering, and so she looks at root-raveled mud and imagines how the laws of physics play out. "The roots stabilize all this muck. But the leaves are just as important." She stoops to wave a floppy blade in her hand. "All this foliage will slow water currents. Stuff will drop out of the water and pile up." I later read that seagrasses are one of the Earth's most effective "bioengineers," building and reshaping habitats through their growth. They don't just live in a habitat; they bring more habitat into being, homemaking. Scientists call this process a "biogeomorphic positive feedback loop," life as a geologic force.

At Blackness, eelgrass hastened the burial of the harbor. Assuming that the harbor was at least a couple of meters deep in the 1700s, one-quarter million cubic meters of mud piled up in the western bay alone. That's twenty-five thousand dump truck loads. Seagrass is so good at sediment gathering that archaeologists call it a vault. In the Mediterranean and off the Australian coast, seagrass caps and preserves shipwrecks. On the Danish coast, Mesolithic and Neolithic villages and fishing weirs are enclosed under eelgrass. No one has excavated the mud at Blackness, but surely eelgrass encloses remnants of busier times.

In the short term, seagrass meadows can accumulate as much as three centimeters of sediment per year compared with unvegetated mud. They literally build up the sea floor by trapping material in their meadows. Over the longer term, accounting for erosion and storms, rates are lower, a few millimeters per year, but over decades this accretion can stave off the loss of coastline from sea-level rise. When storms beat against the coast, seagrass beds absorb the blows, shielding waters and land behind. Seagrasses are champion builders, as are other coastal flowering plants such as saltmarsh grasses and mangroves.

In many places, seagrass meadows are so extensive that they've rebuilt the coastline. In Gathaagudu, also known as Shark Bay, Western Australia, seagrasses protect a 23,000-square-kilometer bay, creating a sanctuary for sea mammals, fish, and rare microbial mats called stromatolites. A dozen seagrass species live at Gathaagudu, each specializing in a different kind of water depth, salinity, and exposure. One of these, a single plant of *Posidonia australis*, is the largest known living creature on Earth. The plant is 180 kilometers across at its widest point and grows from stems that spread out from a single seed that germinated about 4,500 years ago. In the western Mediterranean, seagrasses are even older, albeit smaller. Off the coast of the Balearic Islands, Spain, one fifteen-kilometer-wide clone of the seagrass *Posidonia oceanica*, Neptune grass, is at least 100,000 years old, perhaps the oldest living individual of any species on the planet.

Seagrasses do more than lift the seafloor and protect the coasts; they bury and store carbon in their sediments. Dead plankton, sediment particles, and scraps of plant material drift in, then get trapped. The prodigious growth of seagrass adds to the accumulation. The leaves of some seagrass species grow by a centimeter each day. Seagrass can be more productive than the most well-tended agricultural field.

Worldwide, seagrasses bury about fifty million to one hundred million tons of carbon every year. Along with other coastal flowering plants, mangroves and saltmarsh grasses, seagrasses yearly capture as much carbon in their sediments as all forests put together. Once captured, carbon in seagrass meadows can be trapped for millennia. In contrast, much carbon in forests stays locked up for mere decades. Yet these seawater flowering plants occupy just 3 percent of the total area of forests. The sea plants are champion carbon buriers because they lock up carbon thirty to fifty times faster than the lushest woodland or rainforest. Some of the benefits of this stored carbon are offset by methane, a powerful greenhouse gas released by the growing seagrasses, but the net effect on the climate is still positive. Although seagrasses, mangroves, and salt marshes occupy only 0.5 percent of the ocean, they account for over half of all the carbon taken out of the atmosphere by the oceans each year. Small but mighty, these "blue carbon ecosystems" play an oversized role in sucking up some of the excess carbon that we pump into the atmosphere.

As Julia and I potter around on the mud, we find a few bits of plastic caught in the eelgrass: a yogurt container lid, a frayed piece of rope, and some fragments of white and blue hard plastic. Eelgrass captured and held these scraps. Studies of microplastics show that a handful of seagrass mud can contain thousands of microscopic pieces. In the Mediterranean, tangled seagrass roots and plastic fragments break loose as floating sea balls that get blown onto beaches. In this way, plastic is sent back to the land from which it came, the ocean spitting back the insult of plastic pollution.

Julia laughs and points: "Ooo, crabs!" Shore crabs with mud-encrusted carapaces scuttle sideways over the flattened eelgrass stems. They're small, each would fit comfortably in my palm, and have endearing stalked eyes and tiny pincers. As kids, we'd pluck them by the edges of their shells to lift them to peer eye to eye. I'm surprised to see so many. We're just downstream from the Grangemouth oil refineries. Pollution also flows into the Forth from towns and fields. Yet, in the eelgrass, animal life seems to be thriving or at least persisting.

"Here's a razor clam," I say, "and, look, oyster shells. That's amazing." Oysters have been extinct in the Forth since the early twentieth century, annihilated by overharvesting, yet their shells remain. "This shell could be more than a hundred years old. That's pretty cool. Let's leave it."

Other mollusks are doing better here. The mud is dotted with the white shells of cockles and macomas. Mussel shells are abundant, but we find no live ones. Periwinkles and spire shell mudsnails ply the seagrass stems and creep along the mud. Between the seagrass leaves, a few amphipod shrimp poke out of tiny mud holes. We see one ragworm, longer than my hand, slithering in the wet area where water is draining out of the seagrass meadow. Larger holes and mud castings suggest that these and other sea worms are common.

We're getting a glimpse of how seagrass meadows are havens for animals. Many small invertebrates graze on eelgrass leaves, live in the leaf bases, or make their homes in the sediment. The calmer waters and abundant growth of seagrass are like rainforest canopies, lush green homes for thousands of animal species. Fish such as flounders use the meadows as breeding grounds. Pollock youngsters spend the summer hiding and feasting amid the eelgrass stems. Rays, catsharks, gobies, wrasse, pipefish, and dragonets all need eelgrass.

Eelgrass supports birds, too. Its stems and seeds are one of the main foods of grazing wigeon. Half a dozen other duck and goose species, along with cormorants and shags, prefer to feed over eelgrass, nibbling on leaves and diving for fish and clams. In bays closer to the open sea, razorbills, guillemots, and puffins visit eelgrass to feast on fish, especially pipefish.

Elsewhere, in subtropical and tropical waters, many sirenians, or "sea cows," manatees and dugongs, are almost entirely dependent on seagrass. Using a bristly and muscular split upper lip, they grasp and tear the leaves, pushing them into their mouths where pads and back teeth grind the seagrass before swallowing. A one-thousand-kilogram manatee can eat a tenth of its body weight in seagrass in a day. Sirenians evolved from hoofed land mammals about fifty million years ago and have been munching on seagrass ever since. Like grasses on land, seagrasses evolved to be heavily grazed by mammals, resprouting from buried stems and roots. Now sirenian populations have been decimated by hunting, pollution, and injury from ships. Seagrass meadows are, for the first time in millions of years, often devoid of these large grazers.

Green sea turtles tell a similar story of abundance and decline, although this species has, thanks to conservation efforts, lately started to rebound in some parts of its range. Like sirenians, green sea turtles graze tropical and subtropical seagrass meadows, sometimes so voraciously that they strip the greenery down to the sand. In the past, green sea turtles were astonishingly abundant, perhaps numbering in the tens or hundreds of millions. English colonist Edward Long, writing about shipping in the 1600s, reported that "vessels, which have lost their latitude in hazy weather, have steered entirely by the noise which these creatures [green sea turtles] make in swimming, to attain the Cayman isles." Describing Columbus's second voyage in 1494, Andrés Bernáldez reports that the turtles were "so numerous that it seemed that the ships would run aground on them." Even allowing for some exaggeration in these accounts, we're reminded that, before large-scale colonial depredation of sea turtles, tropical seagrasses hosted communities of large grazers that rivaled the bison populations of North America or the herds of mammals on African grasslands. Today, sea turtles so love seagrass that scientists can use them to find previously undocumented meadows. In the Red Sea, green sea turtles tagged with satellite tracking devices guide researchers to seagrass with perfect accuracy. The turtles far outperform computer predictions derived from satellite imagery of where seagrass grows. Diagrams of turtle movements are like treasure maps. Here lies buried carbon.

We don't usually think of the ocean's productivity as being supported by flowering plants. But even though seagrasses are confined to shallow waters, their meadows serve as nurseries for young fish and invertebrates, and feeding places for larger sea animals such as birds, turtles, and mammals. The same is true of mangroves and saltmarsh grasses, two more groups of flowering plants that are essential to healthy oceans. When biologists call the evolution of flowering plants the "terrestrial revolution," they forget that flowers also made homes in the ocean and created opportunities for tens of thousands of animal species.

The bland-looking slimy strands around Julia's and my boots are unsung environmental heroes. Lowly, ignored plants turn out to be saviors. How terrible are our senses when it comes to the sea! Our boots are oozing into one of the most productive and important ecosystems on Earth, but all we see is mere "weed" and "mud." A recent survey of news outlets found that coral reefs receive one hundred times more media attention than seagrasses. Even salt marshes and mangroves get three times the coverage. This despite the fact that, per hectare, seagrasses can provide two or more times more value from carbon storage, water-cleaning ability, and habitat for animals. Lacking the visual grandeur of the mountains, old woodlands, or coral reefs, seagrass is easily overlooked. Even when we notice it, seagrass can look like a rather unexciting seaweed poking from mud, as Julia and I discovered at Blackness, not a flowering plant that has transformed the edge of the oceans. In my own schooling in biology, seagrass was mentioned not once that I remember. To this day, biology textbooks usually highlight rainforests and reefs, and do not mention seagrass. The United Nations Environment Programme calls them a "forgotten ecosystem."

Not everyone has forgotten. Julia and I descend from people who lived on or near the coasts of England, Scotland, and Ireland, all places where seagrass supported vibrant fisheries and shellfish, provided roofing and bedding material, and likely supplied food in the form of harvested seeds, as eelgrass did for many coastal Indigenous cultures in North America. Yet, Julia and I inherited no knowledge about seagrasses. Industrialization of life in Britain severed the intergenerational flow of knowledge. From forebears who knew the sea, we inherit amnesia. I knew more about mythic Mediterranean sea deities than the real seas of the British Isles that nurtured my forebears. Old, imported tales of Poseidon and the Nereids are wonderful, but insufficient to bond us to the living seas of our homes. Contrast this ignorance to the cultures of Melanesia and Micronesia. There, seagrass and its habitats are named in a dozen different ways, each name highlighting a different habitat. Around the village of Naro, Solomon Islands, some meadows are off limits to fishing. In Palau, the shapes of the seagrass flowers are used as inlays in woodwork. In villages around the Roviana Lagoon, New Georgia, people invoke seagrass spirits to help with fishing, and knotted seagrass brings good fortune. For people in Melanesia and Micronesia, seagrass is honored in everyday life as a source of livelihood and a spiritual home.

Now, amid a climate and extinction crisis, we're all dependent on seagrass as a coastal defender, bastion for biodiversity, and carbon store. How to reclaim lost knowledge and connection? Science can help. Measurements of carbon in mud and biodiversity in meadows open our imaginations to the wonders of seagrasses. But alongside the abstractions of carbon tonnage and lists of species, we need sensory delights: wellies squelching through muck, fingers discovering meshed plant stems in ooze, the invigorating aroma of sea mud, and smiles as we watch crab antics. These experiences reorient our bodies, emotions, and minds, individually and collectively as families and communities.

As Julia and I clamber up an eroded seawall, headed back to land, our shared experience has restitched part of a lost connection. Seagrass meadows not only fringe our everyday terrestrial world, but hold it in place.

✽ ✽ ✽ ✽ ✽ ✽

Few transformations in the history of life have been as extreme as the embrace of the ocean by seagrass. Like whales and dolphins, modern seagrasses descend from land-dwelling ancestors. But marine mammals surface for air. Seagrasses often live entirely submerged. Why did they take the plunge, turning their back on the land to become sea creatures? The world's continents are fringed by vast expanses of sand and mud. Until seagrasses came along, no plant or seaweed could grow for long in such shifting, unstable conditions. Seagrasses brought to these sea bottoms buried creeping stems, strong roots, and pliable leaves. Not only could seagrasses tolerate this oozy, muddy habitat, they built more of it.

Alongside their sinewy body form, the ancestors of seagrass carried with them from dry land two useful features that no sea creature possessed. The first is flowers, the second a molecule, lignin. To these useful inheritances from their land ancestors, seagrasses added a suite of innovations that helped them to thrive in the salty, wave-pounded margins of the oceans.

Seagrass flowers are perhaps the most visually underwhelming blooms in the world. When I first encountered them at Blackness in the early summer, I didn't even realize they were flowers. On a later visit, after looking at diagrams online, I bent a few of the thicker eelgrass blades and saw a row of lighter-green chevrons tucked within the blades' folds. That was it. Unimpressive to human eyes, but amazing in seawater currents.

Under a microscope, the elegance of the seagrasses' flowers is more apparent. Although different seagrass species take different forms, all dodge the destructive power of water's drag by miniaturization and streamlining. Water slips right over them. Petals are gone or reduced to scalelike shields. Some tropical species, like paddle grass, have long, threadlike stigmas that finger the water for pollen. Others, like the ribbon weed that grows in the bays of Gathaagudu, Australia, poke their flowers from tiny spikes atop leafless stems. The more violent the shore, the more streamlined the flower. Eelgrass is perhaps the most extreme of all, flowering inside a stem with its sexual parts barely poking out. The lashings of springtime Atlantic storms can do no harm.

The female stigmas and the male pollen of seagrasses use the flow and viscosity of water to their advantage. Tiny eddies form around them, encouraging pollination. Pollen grains are, in most seagrasses, threadlike. When they come near a stigma, instead of drifting past, as would a spherical pollen grain, they turn in the water current and spin toward the stigma. If they miss, a swirl of current swoops them up to the next stigma. Water currents do most of the work of pollination, but in turtle grass, a tropical species, small animals ferry pollen, too.

One of the great advantages of flowers is the maternal care they provide. In seagrass, the fruit and seed are like a bathyscape, a well-provisioned protective capsule in which the embryo can drift and explore. Some seagrasses float for months over hundreds of kilometers; others settle within hours a few meters from their parent. In a few species in the Indian and Pacific Oceans, naked embryos cling to their mother flowers and are nourished through placenta-like connections until the embryos are ready to break off.

Once they start growing, another legacy from land ancestors helps seagrasses. Their buried roots and stems are strong as wire. Their leaves, although flexible, are very hard to tear. This toughness is imparted by lignin. A molecule originally evolved on land to build tree trunks and support leaves, lignin is what gives plant material its strength. Wood is full of it, cotton wool has none. Paper is wood with the lignin digested away. By bringing to the seas this terrestrial invention, seagrasses grew in ways that no other ocean species could, in rugged mats of almost unbreakable fibers. Lignin also resists decay, helping seagrass to build up sediment as it grows.

Flowers and lignin were helpful bequests from seagrass land ancestors. The rest of seagrasses' success in the ocean came from their own innovations in body structure and collaboration.

Air channels run through the interior of seagrass leaves, stems, and roots. These act as self-filling scuba tanks. Oxygen made by the leaves during photosynthesis travels through the channels to the rest of the plant, including roots buried in oxygen-starved mud. Air channels also buoy the leaves, holding them up to the light.

Water reflects and absorbs light, so from a plant's perspective life in the underwater world is like growing in perpetual shade. Seagrasses cram the emerald chloroplasts that do the work of photosynthesis into the very top layer of the leaf. Seagrass light-gathering pigments are reconfigured, each seagrass species finding the pigment palette that best embraces the blue-green watery light of its home.

Seagrasses have transformed, lost, or duplicated dozens of genes. As is true of many revolutions in flowering plants, some seagrass lineages doubled, then edited, entire genomes. Genes for pumping sodium multiplied, to purge excess salt. Growth hormones and genes that respond to stress were reworked, some lost, others boosted. This is especially true for genes that battle herbivores. In the sea, old enemies are gone, but new grazers ply the meadows. Genes to counteract iron deficiency increased because seawater has just 1 percent the iron of fresh water or soil. Genes for making cell walls reverted to a formula found in algae. When Julia and I confused seagrass for green algae at Blackness, we unwittingly glimpsed a genetic truth. Seagrass has succeeded by becoming alga-like in some of its molecules and cells, all while keeping hold of some bonus land-plant traits.

Like orchids, seagrasses owe much of their success to partnerships. Clonal growth is the first community-building exercise. Instead of going it alone like a tree trunk, seagrasses self-multiply. The lush meadows that result make excellent hiding places for small invertebrate animals. These crustaceans, worms, fish, and others keep their home clean by grazing on the algae that encrust seagrass leaves. This incessant nibbling stops algae from smothering the seagrass.

A three-way mutualism among seagrass, hatchet clams, and bacteria detoxifies some meadows. The clams shunt sulfur- and oxygen-rich water to bacteria that live inside the clams' gills. The bacteria alchemize these chemicals into complex food molecules, some of which they send to their hosts, the clams. This chummy relationship lowers sulfur concentrations in the mud, allowing seagrass to thrive, which in turn provides a protective meadow for the clams. Corals, too, have a mutualistic relationship with some seagrasses. In the Caribbean, seagrass meadows give small corals a safe and food-rich home. The corals cluster around the base of seagrass shoots, where they deter excessive grazing by sea turtles. Without their coral sheaths, the seagrass are often eaten down to the sediment. Without the seagrass, the corals would be swept away by currents. Wherever it grows, seagrass builds mutualisms.

Seagrasses are also hubs of microbial activity. The roots of all flowering plants on land export a little sugar to feed bacteria. Seagrasses turn the trickle into a gush. The sediment close to their roots can be as sweet as fruit nectars or tree sap, although to human tongues the mud tastes overwhelmingly of salt and decay. This sugary generosity comes with restrictions. Along with the sugars, seagrass roots suffuse their surroundings with aromatic chemicals, many of which are familiar to us as the aromas of coffee, red wine, and tree bark. These chemicals lock up the sugar. Most bacteria can't benefit from the sweetness around them. But a few have a key to the lock. Among these are bacteria that supply nitrogen to seagrass roots, detoxify sulfur, and use hydrogen as a food. It seems that seagrasses are feeding helpful bacteria while starving others. The plants may also be dumping excess sugars on sunny days, lacing this sweetness with aromatic chemicals to prevent harmful bacterial overgrowth.

Seagrasses feed the bacteria on their leaves and roots. In return, bacteria protect the plants. Like hatchet clams, root bacteria can purge sulfur. In one Mediterranean species, *Posidonia oceanica,* or Neptune grass, bacteria inside roots snare dissolved nitrogen gas and turn it into ammonia and amino acids. The bacteria then shunt these nitrogen-containing molecules to the seagrass, a built-in supply of fertilizer. The seagrass reciprocates with sugars. This symbiosis was only recently uncovered and it is likely that other seagrasses, many of which are known to host nitrogen-supplying bacteria around their roots, have similarly hospitable internal arrangements.

To human senses, fibrous seagrass roots running through sediment can be boring. Just threads and mud. But at the scale of microbes and molecules, the rhizosphere, "root world," roils with life.

No other plant has achieved as complete a transformation of its body, genes, and life cycle as seagrasses. Their success depends on a careful combination of atavism and innovation. They reversed the course of land plant evolution, going back to the seas four hundred million years after their ancestors crept onto land. In doing so, they became alga-like, especially in the structure of their leaves, but invented new flower shapes, air conduits, biochemistry, and ecological partnerships to accommodate the oceans. Remarkably, seagrasses took the plunge many times. All seagrasses descend from a group of flowering plants that includes the arums and water plantains, many of which grow in swamps or along streams. Starting one hundred million years ago, three or four independent groups took to the seas. What we call by a single name, the seagrasses, are, in fact, several different clans that each took to the oceans.

Seagrasses embody adaptability and resilience. This is hope, in biological form. If flowering plants can plunge into the ocean and thrive, surely their creativity and resilience can cope with the upheavals brought by industrialized humans?

Well, no. Or at least, not yet.

❦ ❦ ❦ ❦ ❦ ❦

"You'll be in pairs, with a gun each. The team has been busy loading them while you all were registering this morning." I'm standing with a couple of dozen volunteers in a stiff March wind on the sand flats at Burntisland. The skyline of the volcanic hills and city buildings of Edinburgh are visible to the south, on the other shore across the Firth of Forth. Blackness castle is just out of sight to the southwest.

A fellow volunteer raises their hand and asks, "What did you call that layer earlier? Can you explain what it is?" Lyle Boyle, seagrass officer for the Ecology Centre and project manager for Restoration Forth, answers, "It's the anoxic layer, just a darker layer in the sediment that signifies lower oxygen. The seagrass prefers to live above the anoxic layer, but they can move oxygen down to their roots if they need to."

"So are we planting above this?"

"Yes, that's the idea. Three centimeters deep. We found last year that worked. It's mostly above the anoxic layer."

Lyle turns and points to caulking guns lined up on the shell-strewn wrack line. "We've mixed the seed in and I think we have thirty-two guns ready at the moment. We'll be planting in circles, each one with a radius of a meter. Each person will plant half the circle.

"A team member will be with you this morning and we'll take some other measurements, the anoxic layer and the height of the sediment. There's Katie with Project Seagrass, Esther with a gun in her hand, she's the scientific lead and also works with Project Seagrass, and Eleri, the seagrass officer for the Seabird Centre based in East Lothian, and there's myself."

Esther Thomsen picks up a caulking gun. "So, these are filled with a mixture of seeds and mud. Each delivers 1.5 milliliters per injection. Don't force it too much." She squeezes the trigger and slate-gray mud glops out of the nozzle, belching out of a side hole rather than the pointy tip, which is blocked by a bolt. "I like to put the tip in the sediment, give it a wiggle to clear a space, then inject. We want to inject about three hundred twenty seeds per square meter, so we're using this string and bamboo to mark out the circles around the flags out there in the sand." She points to the expanse of wet sand exposed by today's exceptionally low spring tide.

"Okay, so, how many seeds does each squirt inject?" a volunteer asks.

"Should be about five or six. If you finish your half circle and your cartridge is still full, just go random in the circle and plant the rest of the seeds."

"Team up in pairs!" Lyle calls. I partner with Jake Norton, director of the Firth of Forth Lobster Hatchery. For years, I had a lobster in a tank in my office at the university where I taught. Homer the Lobster was a rescue from a cookout and a favorite of students, so Jake and I bond over lobster tales. Jake also gives me some tips about whale spotting in the Forth. A deep channel for shipping runs close to the north shore where we stand. In winter, humpbacks come close, feeding. As we plunge our caulk guns into the sediment, planting eelgrass seeds, I imagine new meadows giving shelter to tiny lobsters. A few might drift into open water and be sieved and swallowed by a leviathan. Mighty marine mammals from tiny eelgrass seeds grow.

The scant and struggling eelgrass populations in the Forth don't make much seed, so today we're planting seed harvested from the lush eelgrass meadows of Orkney, three hundred kilometers north of the Forth. Orkney and Forth eelgrasses belong to the same genetic subpopulation, slightly different from the eelgrass on Scotland's west coast. After picking bucketloads of eelgrass stalks loaded with ripe seeds, conservationists stored the harvest over the winter in saltwater tanks. The stalks rotted away, releasing the seeds that are now loaded into our caulk guns. Poking a seed into the mud is easy. Preparation for this moment takes months of work.

Restoration Forth, a project coordinated by the World Wildlife Fund with nine local partner organizations, aims to boost seagrass populations in the Forth. In parallel, they are reintroducing native European flat oysters. No one knows exactly how much seagrass used to grow in the Forth, perhaps ten times as much as today. Because the water is clouded with runoff and algae, light cannot penetrate far into the Forth's water and so seagrass here now mostly grows only in very shallow intertidal water and cannot survive fully submerged in meter-deep water as it does elsewhere. Native oysters are in worse shape, locally extinct, although they used to grow in thick beds throughout the Forth. The people who run Restoration Forth are starting relatively small—four hectares of seagrass and thirty thousand oysters—and using these first attempts to find which restoration methods work best.

The dire loss of seagrass in the Firth of Forth is sad but not exceptional. In the early twentieth century, the global area of seagrass declined by about 1 percent per year. By the end of the century, the annual rate accelerated to 7 percent. In many places, we are destroying seagrass meadows one hundred times faster than they can grow back. The benefits of seagrass—carbon storage, coastline protection, nurseries and homes for countless marine species—disappear with lost meadows. Worldwide, seagrasses are among the most threatened of all habitats.

The causes of these losses are many. Dredging and damage from shipping uproot the meadows. Pollution from agriculture, wastewater plants, fish farms, and industry trigger smothering blooms of algae. Silt clouds the water, starving the seagrass of light. Hurricanes bite away huge chunks of seagrass, especially along coasts where sea-level rise and erosion are severe. Eelgrass was decimated all across the northern hemisphere by a slime mold epidemic in the 1930s, a disease helped along by the weakened state of the plants in polluted waters. Extreme heat is lately also killing seagrass. In Gathaagudu, Shark Bay, for example, marine heat waves between 2010 and 2014 killed over one thousand square kilometers of dense seagrass meadows. When meadows die, the carbon they stored is released back into the water, where some of it flows to the atmosphere as carbon dioxide. Carbon sinks become carbon sources.

As we industrialize, we humans get intolerant of nuanced margins, especially when they are muddy and smell of sulfur. We want sharp boundaries, not subtle edges. Like other coastal vegetation, mangroves and salt marsh, seagrasses get squeezed out as gently sloping shores are replaced with severe coastal walls. This transition is especially evident along the Forth. Land "reclamation" projects built concrete and stone walls, backfilling so that housing, industry, and agriculture can extend to the very edge of the water. On the seaward side of the wall, storms beat away the sediment, yielding a bathtub-like drop into deeper water instead of a complex patchwork of sediment flats of different heights and consistencies. In ecology as in culture and politics, we're replacing diverse gradations with binaries.

The shore at Burntisland is partly shielded from waves by two rocky promontories. The small bay is home to one of the few sandy beaches along the Forth. Despite the often chilly weather, holiday cottages flock the slopes above the bay. At low tide, wet sand and fine mud extend several hundred meters into the Forth and are, in a few places, dotted with small patches of eelgrass.

Jake and I take turns with the caulk guns, squirting muddy slurry into the sand. We can't see the seeds, but Naomi Arnold from World Wildlife Fund earlier showed us some floating in a plastic tube. Each was about the size of a lentil, swollen and bobbing in the tube's water. A few had tiny roots poking out. They're ready.

It seems impossible that poking seeds into wet sand could yield any new plants. I'm used to coddling seeds, planting them in carefully tilled and protected garden beds or indoor seed trays. The suck and punch of moving water will return with the high tide this afternoon. Storms pass through every week or so. Are we being impossibly hopeful?

Perhaps, yes. Last year's seagrass planting here was churned up by storms powerful enough to break down stone seawalls. Seagrass seeds in wet sand had little chance of staying put. Almost none germinated where they were injected, but 5 percent popped up several meters away. Those planted in more protected bays did better, with 12 percent germinating, a rate that compares well with studies of eelgrass restoration in other locations. But even if last year's Burntisland seeds washed into deep water, they helped the project nonetheless. Experimenting with different locations and planting techniques is an essential part of successful restoration. This year's planting depth and locations have been picked based on the most successful methods from last year.

In another sense, my discomfort at planting seeds in such a seemingly impossibly goopy, wave-tossed environment reflects a lack of imagination on my part. For one hundred million years, seagrasses have been sowing themselves into exactly this kind of place. They've evolved to drift away and take root. The seeds I'm used to at home—tomato, lettuce, basil—have lately evolved to need human care. Scattering them to the "wild" would not work. I realize that my expectations of what a seed can do have been domesticated. For a seagrass, a salty, wave-tugged sand flat is a wonderful seedbed. Sheer numbers help, too. One hectare of a healthy eelgrass meadow can produce ten million seeds in a year. Most of these never reach a suitable growing place. Our caulk guns stack the seed lottery in the plants' favor, bypassing the risky dispersal stage and dropping them into ideal spots for germination.

Restoration projects in other parts of the world show that, given a chance, seagrass can bounce back from devastation. The Wadden Sea, which extends from the Netherlands to Denmark, encompasses the world's largest expanse of intertidal mud- and sand flats. As late as 1900, hundreds of square kilometers of seagrass meadows grew to depths of ten meters in clear waters. Then, runoff from agriculture, cities, and industry clouded the water with algae and silt. Disease spread, and damming and dredging degraded the sediment. Seagrass declined by about 90 percent through the twentieth century. Now, thanks to replanting efforts like those in the Forth, using either injected seed or transplanted adult plants, seagrass is expanding in some parts of the Wadden Sea, although bottom trawling and heat waves are new threats. The same is true in other parts of Europe where seagrass coverage has, in some areas, edged higher after decades of decline. Likewise, in coastal Virginia, eelgrass was wiped out by disease and storms. Then, starting in 2001, a consortium of universities, research institutes, and conservation groups sowed 74.5 million seeds into five hundred plots across a total restoration area of 213 hectares. The project was so successful that eelgrass self-seeded and spread to nearly 4,000 hectares, still only one third of its former extent, but a vast increase from near total absence in the late twentieth century. In all these places, as seagrass returned, water quality increased, sediment started to build, and fish, scallops, and other animal life rebounded.

Seagrass may be among the most threatened of all plant communities, but millions of years of life on the ocean margins taught it resilience. Given a chance, seagrass will flourish. Our role, then, is not to *restore*, a verb that implies that we work on a passive object like a restored piece of furniture. Perhaps instead we *revive*, reawakening the life force of a struggling plant community? I prefer the verbs *kindle* and *spark*. *Kindle* implies careful preparatory work, an intentional gathering and arrangement of materials to ensure that the sparked flame can catch and become self-sustaining. At Burntisland, we volunteers were invited to be part of the exciting work of striking sparks by sowing seeds. The staff of the conservation groups had applied for permits, researched best practices, secured funding, worked with the owners of sand flats, forged ties in the community, and done all the other unsung but essential kindling that allows underwater green fire to take hold.

As we plant, a stinging fine rain blows in from the west. The boundaries between air, water, and sediment are slight. Several people comment on the *dreich* day, a Scots word for drawn-out gray, damp weather. We breathe the briny, mud-wet air and plunge seeds into oozing sediment. Then, dreich turns to silver as the rain ends. The sky's weak light reflects in dozens of pewter and pearly hues from the sheen of the sand flats. When our caulk guns are empty, we slosh through tide pools, back to the staging area in the athletic center car park. From there, we carpool to the community center, a low, whitewashed building on the headland at the east end of the beach.

I expect only tea and biscuits, but discover that this informal gathering is as important to seagrass as poking seeds into the sand.

People have lived along the Firth of Forth for at least the last ten thousand years. For all that time, human and marine life have been tightly interwoven. After the glaciers retreated at the end of the last ice age, people thrived here thanks to a rich diet of shellfish, ducks, and fish, helped by abundant hazelnut forests that grew along the coasts and supplied food and firewood. More recently, fisheries and oysters from the Forth and the open sea beyond sustained the local population. In the eighteenth and nineteenth centuries, oyster and fish vendors, often "fishwives" in bold red-and-white-striped dresses, lugged on their backs seafood to sell on the streets of Edinburgh. For centuries, ports and shipyards along the Forth supported local industries and commerce. The importance of the Forth was evident to all, even to people who lived in town. Now many of us have lost that connection. Jobs are inland. Food arrives through supermarkets. When we buy fish and chips or serve fish for dinner, nine tenths of the fish on British plates is imported.

The local coast is scenic, but seems not to be a central source of life and livelihood.

At the community center, I hear old connections reawakening. Most of us volunteers had never heard of seagrass and were only dimly aware of oyster beds in the Forth. Now we not only know about them, but have bodily experience of working with seagrass. The vitality of the Forth turned from an abstraction, a vaguely remembered lesson from school or a museum, into knowledge lodged in the experience of our hands and senses.

What we learn through bodily experience has a habit of spreading. We tell others. The people here are especially good at sharing. Anyone who shows up on a dreich morning to poke seeds into cold mud is likely an energetic, well-connected catalyst in their community. Gathered around the oversized thermoses of hot tea and coffee are youth group leaders, scuba enthusiasts, park rangers, and school volunteers. I imagine the seagrass stories spreading out from here. With the right kindling, fire builds on itself. I experienced the same sense of lively potential among the volunteers at Panola Mountain, a diverse collection of people active in their communities. There, sowing native grass seed and viewing burning prairies are experiences that spread through human social networks, rebuilding interconnections between plants and people.

Holding my hot mug of tea, I finally understand why nonprofits like Restoration Forth and Birds Georgia put so much emphasis on engaging with volunteers and community groups. If seagrass and oysters are to come back, they'll need a change in human imagination and attention. A couple of experienced field ecologists could have planted the seagrass faster and with less hassle than two dozen volunteers. Yet, sowing human imaginations with the idea and experience of seagrass is just as important as getting seeds into the sediment. If culture and policies value local ecologies and restore health to the Forth, seagrass will once again be able to replant itself.

As we were packing up the caulking guns, I asked Addie Dinsmore from the World Wildlife Fund about her own connection to the Forth. She grew up here and lives in a town just inland. "I didn't know about seagrass until I was part of this project. The more you learn about it, the more you see how much we've lost and what that means for habitat and potential for carbon capture. They're underwater rainforests. Here and in tropical areas.

"When I was wee, the water quality here was bad. I went for a swim and came out with a rash. The pollution was horrendous." She shakes her head as she says this, eyes wide. "Things have improved so much. Now people can go scuba diving. It's very different from how it was.

"People around here care pretty deeply about the Forth as you can see by who's getting involved. There's still a strong cultural connection. That's why the comms stuff [communications and public outreach] we're doing as part of this project is so important."

She turns, smiling, to the gleaming, gray waters of the Forth. "Yeah, I think seagrass is having a moment."

❧ ❧ ❧ ❧ ❧

In the long view, seagrasses' moment began about fifty-five million years ago, in a time period known to scientists as the Eocene. Back then, the tropics extended from pole to pole. Ice did not exist, anywhere. Sea level was one hundred or more meters higher than it now is. Seas covered not only the margins of many continents, but cut across interiors. Although many seagrasses have older pedigrees, it was in the warm, shallow margins of Eocene seas that they first became globally abundant. To this day, many seagrasses live only in warmer waters. Those, like eelgrass, that now live in colder areas descend from warm-water ancestors.

A hothouse Earth was seagrasses' cradle. Now the Earth is much cooler, partly because carbon dioxide levels have fallen from their highs in the Eocene. It is sadly ironic, then, that seagrasses are now protecting us from global heating and its effects. By burying carbon and protecting coastlines, seagrasses shield us from the worst consequences of our fossil-fuel-burning folly.

But shielding has its limits. Even if we limit warming to 2°C, a modest goal given current trajectories, seas will rise about two to six meters in the next two thousand years. A rise of two meters will globally inundate over one million square kilometers of land, an area four times the entire land surface of the United Kingdom or twice the area of California. What will grow in the extensive flooded continental margins? Not much, if we keep tearing up the seafloor and using the oceans as a dump. But if the water is relatively unpolluted and the dredgers, trawlers, and boat anchors stay away, seagrass will expand into these newly flooded continental margins. If so, seagrass meadows will entomb the remains of coastal human civilizations, just as they have buried the bay at Blackness, shipwrecks in the Mediterranean, and ancient villages in Denmark.

Looking far ahead, tens of millions of years into the future when humans have evolved into something else or gone extinct, seagrasses might envelop and revive the coasts that we've abandoned. These post-human meadows will surely abound with sea creatures, lively memorials for what is buried below.

In no way do these speculative futures absolve us of our present-day responsibilities. Visions of future healing should spur us. Seagrasses offer us green vitality in the present. They can steer us away from dire futures for our species and other life on the planet. Weirdly, for the first time in their long evolution, the generative and restorative capacity of seagrass now depends on the hearts and minds of land-based apes. It is up to us to say yes, to open our senses, imagination, and cooperative spirit to these ancient, life-giving flowers.

That humans would meld our bodies and minds with flowers seems at first fanciful and unrealistic. Surely we are too tied up in our own concerns to draw flowering plants into our everyday experience and economy? We've done it before. Consider the lilies, the jasmine, the roses . . .

Rose

Aroma knits

interspecies community

I SIT ON A RICKETY METAL CHAIR UNDER AN OLD
olive tree. It's late morning in May. I'm swigging on my water bottle after ambling under a strong Mediterranean sun. Around the
tree, grasses have grown knee high and are drying to hay. My nose en-
joys the calming aroma of the grass, a respite after more energetic nasal
exertions. The last three hours were an olfactory adventure, a walk-and-
sniff through the International Perfume Museum Gardens near Grasse
in southern France.

Oh, heaven! Hundreds of scented plant species in just two hectares.
Every few steps brings another delight. The gardeners here have
arranged plants by aromatic theme—citrus, jasmine, violet, wood,
spice—but the garden also gives the impression of being partly wild.
The caretakers prune with a light hand. They mulch the soil rather than
leaving it bare. They mow walkways, but leave edges to grow into fountains of uncut grass and wildflowers. Signage in the garden acknowl-
edges unplanned sprouts, "weeds," as important parts of the garden's
diversity. No pesticides are used.

Christophe Mège, the head gardener, explained that alongside developing the plant collections, the intention for the garden is to make it
"lush and generous." That generosity extends well beyond human senses.
Bees, hoverflies, and birds are everywhere. Christophe and his team of
gardeners achieve this lush generosity by composting leaves, bringing in
chopped wood from tree-pruning companies, and spreading horse and
sheep manure from local farms. Trees provide shade for more delicate

plants and habitat for animals. Dead wood on the ground provides nesting sites for local insects.

As part of a management plan that aims to integrate into the site's ecology, Christophe and his team encourage the growth of local plants alongside the focal aromatic plants. So, even in this place where species from around the world are gathered, local plants are foundational, seeding themselves and finding their way with some help from gardeners. The result is the antithesis of formal gardens, places where we force geometric order onto plants, aiming to please only the human eye. Here, humans and plants entwine their work, producing a place full of life and delight for all the senses.

Building a thriving garden is challenging, Christophe told me, in an environment where soil is "parched" in summer and "gorged with water" in winter. His work gives a glimpse into the centuries of hard labor that growers in this region have devoted to the cultivation of aromatic flowers and plants.

Christine Saillard, the director of public engagement and cultural programming for the museums in the Grasse region, told me that the ecological management of the gardens is "difficult for some of our visitors to accept." People arrive with "the idea of a traditional French garden in mind: a well-watered, meticulously maintained, and well-manicured garden with lush greenery." But here, the gardens "preserve wildlife and plant life while using minimal water in a region that is very dry in the summer." A few visitors might struggle with the lack of formality, but the rich bird life and ecology of the site have been recognized with awards from national conservation groups.

During my visit, *Rose de Mai*, May roses, are in full bloom and I spend an hour dipping my nose into their pink petal ruffs. Honeybees and huge violet carpenter bees also make their way down the rows of rose bushes, from bloom to bloom, gathering pollen. The flowers' aroma is a brocade, soft sweetness textured by threads of golden spice and a few strands of musk. This rose variety, also called *centifolia* for its profuse petals and leaves, is a hybrid of European and Asian wild roses. The flowers are so deliciously aromatic that the variety and its ancestors have been grown for centuries as a source of perfume. If you've enjoyed a rosy aroma in a perfume, you most likely smelled the May rose or its close relative the Damask rose.

Next to the rows of rose bushes, I find a patch of pale blue iris coming into bloom, their flowers smelling faintly powdery. Perfumers extract a potent scent from their buried stems. I rub sage leaves between my fingers as I amble toward the narcissus bed, its blooms now maturing to fat pods. Walking uphill, I smell the small lavender meadow before I see it. Mixed with thyme, lavender is the signature aroma of these hills of Provence, a lively, slightly astringent, herbal smell, like pine resin dressed up for Saturday night. Woody jasmine vines sprawl in rows across a path from the lavender. Their flowers will not emerge until later in the summer and, for now, the plants look like feral vineyards.

Resting in the shade of the olive tree, I nestle among less well-known aromatic plants. A grove of bitter oranges yields scented flowers and also aromatic oil from leaves and fruits. Sweet flag, a plant with upright spiky leaves, grows in the margins of a pond a few meters away. Its underwater stems are loaded with earthy, camphor-like aromas. Acacias, used for their scented twigs, canopy a path between garden beds. Vetiver grass, a shaggy bunchgrass, is used for its roots.

I especially enjoy a nook called "The Garden of Forgotten Perfume Plants," home to species that were once important in perfumery but are now seldom used. The aromas of some, like figs, have toxic effects and can no longer be legally used in perfumes. Most others were dethroned by synthetic substitutes. Nineteenth-century chemists, for example, discovered how to synthesize chemicals that smelled like violets and lilies of the valley. These floral aromas are hard or impossible to extract, but synthetics offer good approximations. Likewise, hyacinth flowers have a wonderful green, cut-grass aroma and were widely grown for perfumery in the 1920s. But the yield of perfume from the plants was low and now chemists make passable replicas in the lab. Here in the garden, our noses can meet aromas directly from the plants.

This garden is a living museum whose intention is to remind us that, for all its artifice and artistry, perfumery is fundamentally about our relationships with plants, especially flowers. Apart from a greenhouse housing the visitor center, the entire garden is uncovered, open year-round to the elements. The fact that so many different species can thrive here reflects the special microclimate of the region. Mountains to the north supply cooling summer breezes and water, both scarce here. The Mediterranean Sea lies fifteen kilometers to the south and buffers from extreme winter cold. Late winter and spring rains encourage early growth. Abundant sunshine and warmth mean that even subtropical species thrive.

It is from this botanical sweet spot that contemporary perfumery evolved, drawing on scent-crafting traditions and plants from around the world. For the last five hundred years, the people of the Grasse region have grown aromatic plants and refined the art, science, industry, and marketing of modern perfumery. In 2018, the United Nations Educational, Scientific and Cultural Organization (UNESCO) recognized the knowledge and practices of the people from the Grasse region as a significant part of the world's intangible cultural heritage. UNESCO noted skills in "the cultivation of perfume plants, the knowledge and processing of natural raw materials, and the art of perfume composition."

Grasse's perfume industry started, improbably, with the stink of tanneries. The town was a center for processing animal hides as early as the twelfth century, importing raw skins from across the Mediterranean. Grasse's leatherworkers, especially glovers, became renowned across Europe. In the sixteenth century, some of these tanners hit on an idea that would change the fate of the region: scented gloves. Oils and powders steeped in floral and other plant aromas sweetened the sharp tang of leather and imparted lingering aromatic delight to hands. The fashion caught on, boosted in the 1530s by a Renaissance influencer, Catherine de' Medici.

By the seventeenth century, gardens of aromatic flowers, fed by the same streams that sluiced tannery waste, grew all around Grasse. These supplied not only glovers but a new kind of artisan, perfumers. By the eighteenth century the perfume market was so well developed that local tanneries closed and the guild of perfumers broke from the glovers. Grasse was thriving, but the new manufactured perfumes were at first affordable only by the rich, especially those at court. King Louis XIV was known as *le doux fleurant*, "the gentle, sweet flowery one," for his love of perfume, and Louis XV's court was known as *la cour parfumée*, "the perfumed court." In those days, masculinity—and what says *entitled maleness* more than being a king?—was supremely floral. The kings grew some of the plants for their perfumes at the palaces near Paris, but Grasse's microclimate and centuries of know-how kept it at the forefront of perfumery production and innovation.

By the late eighteenth century, Grasse's manufacturers sold perfumes to the middle classes. This expansion of the market for perfume was helped by the increasing use of distilled alcohol as a solvent and carrier for floral scent, a practice that continues today in many perfumes. Later in the eighteenth century, despite the violence and upheaval of the revolution, industries in Grasse persisted. When French Crown and Church lands were auctioned off by the new revolutionary governments in the 1790s, much of the land around Grasse was bought by well-to-do merchants. *Liberté, égalité, fraternité . . . parfumé.*

Through the nineteenth and early twentieth centuries, Grasse's industrialization brought perfume to the masses. We can now walk into a department store and buy inexpensive rosy perfume largely thanks to the mass production and marketing of perfume developed in Grasse.

Today, most raw materials for perfumes are synthesized in labs or grown in Eastern Europe, Turkey, southern Asia, and North Africa, but Grasse's roses and jasmines still supply high-end perfume brands.

The foundation of the region's preeminence in perfumery is water. Because tanneries needed water, Grasse grew up around some of the few dependable and abundant streams coming down from the mountains. These water sources later irrigated fields of roses, jasmine, tuberose, and other aromatic plants, species that could not otherwise grow in the region. The main square in Grasse is called Place de la Foux, named for the spring that emerged there. In 1568, the local council implemented a town-wide plan for the distribution and management of water. This scheme remained in places for centuries and made sure that gardens and fields were well supplied and not polluted by tanneries.

In the nineteenth century, canal builders vastly increased the supply of water to Grasse. First, they diverted part of the Siagne River, carrying its waters from a gorge twenty kilometers to the west via Grasse to Cannes. This canal spurred growth of tourism on the coast and helped inland flower growers. Then workers carved and tunneled through mountains to bring waters from the Foulon spring from the east. By the 1890s, the roots of Grasse's plants were kept moist by water diverted from a large swath of the surrounding mountains. This is as close to perfect as you can get for roses and many other aromatic plants: never-ending water for roots, abundant sunlight for leaves, and a moderate climate.

Sitting under the olive tree, enjoying the garden, it is tempting to imagine myself in a lightly tended natural space. But the botanical abundance here is a product of nineteenth-century engineering. My bench is a few meters from the Siagne canal, which runs out of sight, upslope. The waterway irrigates the garden and supplies the garden's frog-filled ponds. When I swig from my water bottle, I'm drinking from the diverted Foulon spring, which, to this day, supplies Grasse with drinking water. Before the canals, before the tanneries, people here grew wheat, fed by spring rains, alongside drought-resilient olives and vines. Perfumery changed these old ways around Grasse, replacing thousands of years of grain, oil, and wine cultivation with crops that fed the nose.

By the late nineteenth and early twentieth centuries, two thousand hectares of fields devoted to aromatic plants grew near Grasse. During the flower-picking season, May and June for roses and July into October for jasmine, local farm laborers were joined by workers from Italy and North Africa. Yearly, they picked two billion or more roses, each flower plucked by hand within a few hours of dawn, then hauled to nearby distilleries. Jasmine flowers are smaller and their vine tresses trail flowers to the ground. Field workers harvested upward of ten billion jasmine flowers every year. They called them *parfum de mal au dos*, the "backache perfume." The ache continued into the autumn when the vines required hilling up with soil or mulch to protect them from winter's chill. Rose and jasmine gave the most abundant harvest, but the fields here also yielded hundreds of tons of violet leaves, tuberose flowers, oranges, narcissus, and mimosas. All through the growing season, carts loaded high with sacks of perfumed flowers and leaves rolled through the streets of Grasse.

The aim of all this labor was to capture and concentrate aromas. A little perfume goes a long way. One drop can scent our skin or clothes for days. Such potency comes from extreme concentration. During the extraction process, two tons of roses make one kilogram of extract. A tiny drop of rose oil contains the aroma of thirty roses. No wonder we smell it across the room or in the lingering perfume of a stranger passing on the street. Like hard liquor and plant-based pharmaceuticals, perfumes gain their power through focus. What the plant made in moderation, we amass.

To collect the aromas of roses, the perfumers of Grasse used distillation, a method still used here as well as in Bulgaria and Turkey, where rose production is an important industry. Workers shovel mountains of pink petals into heated cauldrons. Giant copper hoods capture the steam and shunt it through pipes. The steam that cools to liquid in the pipework is the aromatic essence of rose, which then drips from pipe ends. In the cauldrons, cooked petals are reduced to a yellow-brown mess. The bottled essence is, in some years, worth more than its weight in gold.

The airy, sweet aroma of jasmine is too delicate for distillation. Heat kills the scent. Instead, factory workers in Grasse, almost all local women, pressed individual flowers into sheets of rendered, odorless animal fat, usually lard or tallow. The fat was smeared in thin layers over glass sheets in wooden frames. Each frame held up to one hundred flowers squashed facedown in the ooze. The jasmine flowers' aromas seeped from petals into fat. After a day or two, the women unpicked the spent flowers and pressed new blooms to the sheet. After several rounds, they scraped the fat into vats where washes of alcohol captured and further concentrated the scent into an absolute, an extract so dense with aromatic oils that it is viscous and dark. The leftover fat retained some aroma and was used to make soaps. This process of room temperature "enfleurage," literally enflowering, was created and perfected in Grasse and carried the jasmine's aroma unsullied into the perfumers' bottles. Later in the nineteenth century, Grasse's perfumers invented a process that used a solvent, hexane, to capture the aroma without the labor-intensive process of enfleurage. Both methods were used in Grasse into the early twentieth century. Today, hexane and other solvents are still used to extract jasmine's aromas. The glass sheets and wooden frames used in enfleurage are stacked in museums.

These methods of scent extraction have ancient roots. By 3500 BCE, Sumerians used capped ceramic vessels to distill essences from herbs, a technology also invented in China by 2000 BCE. These early vessels were later improved by the addition of cooling pipes and distillation heads, and the use of glass, allowing more precise control of the extraction process. By the fourteenth century in Syria, rose water was distilled using stills with multiple heads leading from a central heated pot. Enfleurage, too, was a refinement of an old process. By bathing flowers and other plant parts in oil or fat, ancient Chinese, Egyptian, Indian, and Persian cultures all developed scented unguents.

Extraction methods seek not only to concentrate but to contain. Aromas are what chemists call volatiles, lightweight molecules that drift away in the breeze. Most are soluble in fats and oils, so they easily slip through the cells' fatty membranes and out into the air. Flowers smell good only as long as they keep making new oils to replace those released by evaporation. The art and science of perfumery takes what is ephemeral and makes it long-lasting, storing volatiles so that they can be delivered at will. In the garden, we enjoy the scent of a May rose for a few weeks each year, and even then the aroma is best only in the morning hours. In a bottle, May rose is perennial, ready to bloom on our skin whenever we desire.

❦ ❦ ❦ ❦ ❦ ❦

On the May roses in the museum's garden near Grasse, honeybees dive into the bowls of petals and run frenzied circles, rummaging through the central mop of pollen-bearing anthers. With swift strokes of all six legs, they ball the pollen into baskets on their hind legs, food for hungry larvae back in the hive. As they move, the bees oil themselves in aroma, the ultimate rose-petal bath.

The upper surfaces of rose petals are entirely paved with conical cells swollen with oily aromatic volatiles and pink pigment. Cells on the lower petal surfaces, too, are loaded with scented oils. The volatiles pass right through the cell membranes and out into the air. Every petal cell is a living aroma diffuser.

Bees sense rose volatiles through hundreds of minute hairs that fur each antenna, along with odor-sensitive pits between the hairs. Scrabbling through a flower must be an intense aromatic thrill for the bees, every sensor slathered. This experience feeds the honeybees' astonishing skills of aromatic discrimination and memory. They can tell the difference between hundreds of floral scent variants and learn to associate nectar with aroma after a single visit to a flower. They remember especially rewarding aromas for the rest of their six-week-long lives.

Giving pollinators a memorable hit of aroma is why flowers make scent. Insects learn to associate floral aromas with rewards much faster than they do with colors. Aromas also come in thousands of variants. Each flower species crafts its own aromatic aide-mémoire. If you're a flower offering costly pollen and nectar to insects, you had better make sure the visitors remember exactly who their beneficiary is. This ensures repeated visits and fidelity. There are no anonymous donors in the flower world. Aromas boldly declare identity.

For plants, aromas are a language. Each chemical variant is like a word, a building block that, in combination with other words, creates and communicates relationships and meanings. Flowers are a latecomer to this chemical chatter. Volatiles first evolved to unite the cells inside a plant into a communicative network, coordinating the growth of tissues or the deployment of chemical defenses, just as nerves and hormones do within human bodies. Because volatiles slip easily through membranes, they zip signals from one cell to another. Volatiles come in thousands of forms and so the signals are precise. A few volatiles evolved not only to link cells within individual plants, they also waft through the air to other plants, keeping neighbors apprised of the latest plant gossip: How's the soil today, what's biting you, is it time to bloom?

The aromatic language of plants also signals to sap- and leaf-feeding insects. Go away, say the chemicals, this plant is nasty. These volatiles not only taste bad or are poisonous but advertise the plants' unpalatability. A few plants extend the conversation to the enemies of their enemies. When attacked by insects, plants release volatiles into the air and soil that summon parasitic wasps, predatory mites, and insect-killing nematode worms. The bodyguards, drawn in by plants' scent, kill the plant-eating pests.

As flowers evolved and diversified, they extended the language of plant volatiles in revolutionary ways. Instead of whispering among themselves, threatening their enemies, and shouting for help, flowers used volatile molecules to knit themselves into cooperative unions with animals. A few, like many orchids, later turned cooperation into exploitative illusion. But for most flowers, volatiles are plainspoken statements of the flowers' intention to work with their pollinators. Here is food, they say, please come help yourself.

Some flowers speak to animals with great specificity, signaling to just one or a small number of pollinators. The sexually deceptive orchids that we encountered earlier deploy unique blends of volatiles, attractive to only the males of a small number of insect species. Other flowers also focus their aromas on single pollinators. Flowers of the drooping fig, for example, a species that lives in South Asian forests, uses a single volatile chemical to attract the attention of the females of one species of wasp. This insect not only pollinates the flower, but also lays her eggs inside. The tight mutualism between fig plant and wasp is only possible because the fig's unique aroma is a private scent beacon cutting through the confusion of the forest, an odor plume that allows wasps to find their flowers.

Aromas offered for a single pollinator species are the exception, though. The dozens, and sometimes hundreds, of volatiles in a flower's aroma typically invite broad classes of pollinators. Bat-pollinated flowers, like those of cacti in North American deserts and some tropical rainforest trees, tend to have a sulfurous, fruity whiff. These flowers often also have reflective cups to bounce back the echolocating sounds of bats, the acoustic equivalent of a brightly colored bloom. Flowers pollinated by birds have little to no scent, relying instead on visual signals like bright red petals. Fly-pollinated flowers like hawthorn have sweaty, sharp-smelling nitrogen-containing molecules mixed with sweeter notes. A few, like the arums, summon flies and beetles with ammonia and sulfurous molecules that mimic carrion or feces. Moths and butterflies are especially attracted to volatiles known as benzenoids, many of which smell to us intensely sweet. Bees and wasps like these smells, too, as well as terpenoids, another class of volatile that includes aromas like eucalyptus, menthol, cloves, and lemon.

Within these broad classes, the flowers of different plant species, even those that are closely related, often have unique aromatic blends. This contrasts with the chemical conversations inside cells. There, most chemicals are similar or identical among species. Rose, fig, and orchid all use much the same molecules to make new DNA, capture sunlight, and coordinate growth among new cells. But each flower uses different volatile chemicals and different blends among them to speak with pollinators. This variety of aromatic expressivity is created by the richness of ecological relationships.

Wild *Narcissus* species in Spain, for example, make more than eighty different floral aroma molecules. A few, like citrusy limonene and earthy myrcene, are shared among all species. But most are idiosyncratic, and every species has its own blend. A pattern is discernible within this diversity. Species whose primary pollinators are moths and butterflies—species like paperwhite narcissus, rush-leaf jonquil, and autumn narcissus—have aromas dominated by indole, a musky, almost fecal aroma, and other chemicals with a leafy, spicy aroma. Those pollinated mostly by bees and flies—petticoat daffodil, angel's tears daffodil, and the endangered *Narcissus bujei*—have aroma profiles that are mostly fruity and sweet.

The wide range of floral aromas is crafted by genes that snip and glue molecules. These genes descended from genes that chop up and rearrange food and waste in plant cells. Through gene duplication, these workaday genes made copies of themselves that then evolved into aroma-makers. The duplicates concocted a language of scents that waft out of cells, connecting to plant and animal neighbors. This language now includes thousands of aromatic "words," each delivering a unique message.

When I sniff the aroma of roses in the Grasse garden, for example, I'm inhaling four hundred different volatiles. The smell that we call by one name, "rose," is a conviviality of molecules. One of the most important rosy aromas is geraniol, a volatile oil. In rose petals, geraniol is shaped by a variant of a housekeeping gene found in almost all living beings. The gene duplicated many times in the ancestor of wild roses, strawberries, and cinquefoils, a clan of about eight hundred species. After rose ancestors split from the others, more duplication followed, scattering copies of the gene across many chromosomes. During one of the duplications, another chunk of DNA got inserted that now acts as an on-and-off switch for geraniol in rose petals. Today, the rose varieties that most thrill our noses have the most gene duplicates.

The blissful experience of smelling a fragrant rose is the result of genetic exuberance. Rose ancestors went wild with the gene-copier machine, then put the swarm of copies to work making volatile oils.

Flower aromas evolved to speak to pollinators. But enemies also smell flowers. Herbivores with a taste for petal, pollen, ovary, or sap can follow the scent of flowers to food. Flowers are caught in a bind. Too much aroma attracts the wrong kind of attention. Too little and the good pollinators don't notice.

Lab experiments reveal just how knife-edged this balance can be, and how flexible flowers are in response. When field mustard plants are pollinated in the lab by bumblebees, with no herbivores sniffing around, they evolve fragrant flowers that are especially attractive to bumblebees. It takes only six generations for the mustards to evolve into bumblebee thirst traps. But add caterpillars and the flowers put the brakes on floral evolution. They evolve demure flowers. Caterpillar-munched plants must invest in chemical defenses, which takes resources away from the flowers.

The plants are too exhausted to be sexy. The caterpillar-plagued flowers self-pollinate more than those that experience no herbivory. To help with this self-love, male and female parts evolve to snuggle closer to one another within the flower.

The Very Hungry Caterpillar shuts down floral sexiness. This is true outside the lab, too. When experimenters use pesticide to knock out caterpillars and slugs, but not bumblebees, on wild field mustards, it takes just four generations for the plants to evolve flowers with more fragrant blooms. In the western United States, the night-flowering wild coyote tobacco adjusts its blooms according to how many hawkmoth caterpillars are munching its leaves. Hawkmoths are the tobacco's favored pollinator, but if the moths lay too many eggs on the tobacco, what was a helpful relationship turns sour as the caterpillars destroy leaves. Normally, the tobacco blooms at night, with gorgeously aromatic blooms. But if too many caterpillars plague the plants, the tobacco flowers shut off scent. They also switch to blooming during the day. This deters further visits from hawkmoths. Luckily for the tobacco, day-flying hummingbirds pick up some of the slack. The aroma of many flowers is a compromise between sex and safety. Every plant finds its own balance between attracting pollinators and hiding from herbivores.

Sometimes, flower aromas become part of the chemical language among animals, often to the plants' benefit. When a honeybee returns to the hive, the lingering scent of the flower on her body helps to recruit and reorient new forager bees. The returning bee, weighed down with nectar or pollen, communicates her success to others in the darkness of the hive through a dance and scent. The dance, a figure-eight waggle and buzz on the honeycomb, tells the other bees how far and in which direction to fly. As the bee dances, nest mates crowd around and get a whiff of floral scent. The newly recruited forager bees remember the aroma as they fly out of the hive, searching for the flowers. The directional information in the waggle-dance gets them to the general area, but the final part of the search often relies on aroma, guiding the bees to particular flowers amid a jumble of vegetation. When they arrive, the bees are sometimes helped by scent marks left by the original forager bee, little aromatic footprints on the flower. These marks give a thumbs-up or thumbs-down for rewarding or depleted flowers.

By offering abundant and distinctive perfume only when their flowers are at peak ripeness, plants help social insects to recruit nest mates. In the hive, plant language becomes bee language. Sister, where did you get that perfume? I *have* to get some . . . ah, there are the flowers.

In South and Central America, some bees go further, using floral aromas as a love language. We're not the only species to perfume ourselves for romance. Male bees woo females using scents that they've gathered from flowers. These male bees land on petals or other scented parts of flowers and dab drops of scent-absorbing oil onto the flower. Using brushes on their forelegs, the bees then sweep the perfumed oil to the hind legs where combs squeeze it into hairy grooves and pouches. The perfumers in Grasse, it turns out, did not invent enfleurage. Bees got there first. This should not surprise us. Bees eat nothing but pollen and nectar and so are literally made from flowers. That some of their bodies are shaped to scoop up perfume seems fitting.

The presence of so many scent-obsessed bees later affected plant evolution. Today, about six hundred orchid species are pollinated by the perfumer bees. Each flower glues its pollen sacs onto a slightly different part of the bees, allowing focused delivery to flowers of the right kind, even as the perfumer bees visit a range of flower species. The plants that are now pollinated by perfumer bees evolved at least ten million years after the bees. This suggests that, although these bees and flowers each depend on one another, the relationships are not a tight one-to-one dependence like figs pollinated by a single species of fig wasp.

Like human perfume brands trying to distinguish themselves from their competitors, flowering plant species each offer distinctive aroma blends with recognizable profiles. Among orchids pollinated by perfumer bees, for example, flowers draw from the same list of ingredients, just under two hundred aroma molecules. Each flower uses between one and six of these for the dominant "notes" in their floral perfumes, in different combinations and strengths. Where many orchids all grow in the same area, aromatic signatures seem especially distinctive. For the bees, every species has its own taste, a specificity that allows females to reliably find and assess males. Like the plants, it seems that when many bee species are crowded together, aromatic preferences are especially strong and singular.

Picky love languages can lead to rapid evolution. The specificity of bee preferences and flower aromas make for species easily broken into new daughter species. Changes in a single odor-receptor gene in the bright green *Euglossa* perfumer bees in southern Mexico and Central America, for example, split a species. The mutated version of the gene is tuned to different aromas than the original gene, so bees with the new version only mate with one another. For these bees, minor changes in taste for perfume created a new species.

Smell is the most impertinent of the senses. When a bee's antennae tingle with flower aromas or when I sniff a bloom, volatile molecules that minutes ago were inside petal cells bind to animal aroma receptors. Unlike hearing and seeing, senses that have the decency to act through the intermediary of sound or light waves, the sense of smell results from direct physical connection. Through this intimate contact, our senses come into bodily contact with plant genes and their evolution. To smell a flower is to inhale evolution's creative power.

❦ ❦ ❦ ❦ ❦

It is a mystery why humans take such delight in the aromas of flowers. We're not pollinators like bees, so why did evolution give us scent receptors that can pick up floral aromas? Why are our brains wired to find many of these chemicals so enticing? Unlike other attractive smells—good food, the aroma of our mates and children—flower aromas seemingly have little direct bearing on our survival and procreation. Yet, we bring fragrant blooms into our gardens and homes, cultivate and bottle floral scents, and spray the aromas of flowers onto our bodies. Strangely, we know more about floral biochemistry than we do about the origins of our own fascination with the smells of flowers.

It is not all floral aromas that we find pleasing. Flowers like the corpse flower and other arums that mimic carrion or feces are disgusting to us, but irresistible to many flies and beetles. Others, like the orchids that offer love perfumes for male bees, smell to our noses sharp and funky, like mothballs mixed with engine oil, sometimes with a tinge of citrus and buttery vanilla, not the kind of aroma typically sought out by human perfumers. The flowers in the garden at Grasse and in perfume bottles worldwide are a small selection of global diversity in flower aroma. What accounts for the allure of these few?

Our cultures teach us some of the meaning of smells. By watching adults and through years of experience, infants come to understand the significance and valence of aromas. If you grow up in Southeast Asia, durian fruit smells wonderful, but the sulfurous aroma is off-putting to many others. Cheeses and other fermented foods smell delightful to those familiar with them but appalling to people not accustomed to the stench of bacteria- and fungus-matured milk or cabbage. Similar cultural influences affect our reactions to floral aromas. Jasmine and violet were considered masculine scents in medieval Islamic cultures, and jasmine was the scent of lords in Iran, yet men in contemporary northern Europe generally shun jasmine as too feminine. Likewise, roses are considered feminine in most European perfumes, but both masculine and feminine in the Middle East, where rose water has been used since at least Assyrian times, over three thousand years ago. The musky and even fecal aromas present in some European perfumes are considered distasteful in Japan, where the majority of perfume is floral. In the womb, we learn to associate aroma with pleasure. Dissolved in mothers' blood, aroma molecules from vegetables like carrots and seasonings like anise flow through the placenta to fetuses. Once born, infants remember which plant volatiles are rewarding. Whether we learn before birth the aromas of flowers and floral perfumes is so far unknown. If your mother wore rosy perfume when she was pregnant, does your subconscious somehow remember this and give you a special affinity for roses?

The meanings of aromas also change through time within cultures. For example, when Ernest Beaux created Chanel No. 5, using jasmine from Grasse, the perfume was a hit among young fashionistas of the 1920s. Later in the century it seemed dated, an "old lady perfume." Now, the perfume's reputation has been revived as a "classic." Continual change is the point of fashion, allowing those in vogue to signal their social standing, aesthetic acumen, and wealth.

Perfume is a perfect vehicle for fashion culture: Billions of combinations of volatiles are possible—an endless palette for change—and many, but not all, ingredients are expensive, creating a seamless gradation in prices from affordable to outrageous. None of this is new. Writing in the first century, the Roman writer Pliny recounts dozens of changes in aromatic fashions: "Perfume of iris, from Corinth, was long held in the highest esteem, till that of Cyzicus came into fashion . . . after which perfume of œnanthe, from Cyprus . . . and then that of Egypt was preferred . . . then was supplanted by unguent of marjoram, from Cos, which in its turn was superseded by quince blossom." Two thousand years later, he reads like a contemporary fashion reporter, tracking every new release from perfumery houses around the Mediterranean. In Pliny's time, too, some perfumes were cheap enough to be used in soldiers' camps, but others were eye-wateringly expensive, used for "luxurious gratification," pleasures that, unlike jewels and clothes, "die away the very hour they are used."

Not all our attraction to flower aromas is culturally learned, though. Humans worldwide find floral scents enticing. Underneath the ever-changing churn of culture and fashion, a question still lingers. What might be the biological basis of our attraction?

We smell using receptors deep in the nasal cavity, buried in the skull under our eyeballs. There, aroma molecules from the air dissolve into a thin layer of mucus where carrier molecules ferry the aromas to waiting receptor nerves. These nerves bristle with proteins that pick up specific aroma molecules, one protein type per nerve. From the nasal cavity, the nerves run directly to a processing center in our brain. We have about four hundred different receptor types, more than twice as many as honeybees (and far more than the paltry three color receptors in our eyes). Combinations among these are what our brain translates into the conscious experience of aroma. No wonder smell is so multifarious.

We each differ in the sensitivity of our aroma receptors. The experience of smell is therefore highly personal. Quite how many different aromas the average human nose can distinguish is controversial. Some scientific studies claim that humans could theoretically discriminate up to a trillion different aromas, but others say that we can pick out just ten thousand variations. The disagreement stems from just how far we can extrapolate from the results of a small number of carefully controlled laboratory studies. All modern studies agree, though, that the nineteenth-century idea that humans have a poor sense of smell is a myth. Most of us have up to twenty million aroma-sensitive nerves in our noses. In theory, we can sniff traces of almost all volatile molecules. Like all senses, there is wide variation among individuals in the acuity of the sense of smell, partly caused by genetic quirks and partly by nerve damage or degeneration.

The fact that many of us respond so strongly to floral aromas is partly due to the need to find food. At first, this seems preposterous. We might eat an occasional nasturtium bloom in a salad, but we're not a flower-eating species. But deep in our nervous system, our ancestors whisper to us. Floral aromas, for them, could signal food. Before our ancestors evolved into grass apes, they spent about fifty million years as tree-living primates where the smells of flowers and leaves were vitally important guides to food.

Among primates that still live in the forest, flowers are a seasonally important food, and primates from all families except the insect-eating tarsiers munch flowers. White-faced capuchin monkeys in Costa Rica eat more than a dozen different flowers, gorging themselves when favored flowers like night-blooming mallows come into season. In African and American tropical forests, many species of monkey feast on the red, succulent flowers of boarwood trees. These flowers release their pollen in oil, anointing visiting birds to ensure pollination, so the unctuous pollen-bearing anthers make rich meals for monkeys. Chimpanzees and bonobos, our closest living relatives, also occasionally eat flowers. Although they mostly eat fruit and foliage, blooms can make up about 2 percent of their diets. These apes eat flowers even when fruit is abundant, suggesting that they actively seek out floral snacks. We modern humans have lost the taste for eating flowers, but we still have most of the smell receptors that guided our ancestors. Our love of floral perfume is partly a gastronomic atavism.

Floral aromas also guide us to good habitats. Many of the aromatic chemicals in flowers are the same as those found in foliage and fruit, presented in concentrated form by the flower. Linalool, for example, has a spicy, slightly lemony smell and is found in the aromas of many flowers. It is also abundant inside many leaves. Methyl jasmonate is an essential part of the internal communication network of many plants and also a common floral volatile, smelling like jasmine. Our nerves and brain pleasure centers delight in botanical language, perhaps because leafy places are good habitats for humans. A woodland, glade, or meadow filled with vigorously growing plants is more likely to sustain us than deserts, rocky outcrops, or swamps. Aromas guide us home.

Flowers release huge gusts of plant volatiles from a relatively small space. This makes them bewitching, creating illusions of botanical abundance. Sniffing a flower delivers dozens, sometimes hundreds, of aromatic plant molecules in one huff. Inhaling the aromas of a flower, we get microbursts of connection to plants. Manufactured perfumes further ramp up the intensity, packing the aroma of many blooms into a single drop of essence. We don't just bathe in plant aroma, we plunge.

It is perhaps no accident that manufactured perfumes are especially popular in urbanized and industrialized cultures. People worldwide use aromatic plants to perfume the body, yet it is in cities that perfumery trade is briskest. As we lose everyday contact with plants, we hunger for reconnection. Walk into the perfume section of a department store and you're in the aromatic presence of hundreds of plants. Perfume, then, is perhaps an antidote to ecological loneliness. We dab and spritz memories of the forest and meadow onto our bodies, wrapping ourselves in the embrace of green kin. In this view, perfume creates illusion, the scent of healthy flowers and plants where few exist. Or, from a more upbeat perspective, perfumes remind us that, for all our technological sophistication, we are still a species that thrives in community with plants.

A sweet-smelling rose is also a well-defended rose. Floral volatiles not only attract pollinators, they repel nectar-stealing ants, inhibit pathogenic fungi, and weaken bacterial attacks. For humans, this means that plant volatiles can heal us. When death comes, they also delay decay through their antimicrobial effects and cover the smell of rot. The insect- and bacteria-repellent properties of floral aromas were one reason why we drew them into our lives and, sometimes, the afterlife.

Around thirteen thousand years ago, in a cave in what is now Mount Carmel, Israel, mourners lined graves with flowers and leafy vegetation.

The graves had been chiseled out of limestone, then veneered with wet mud. Millennia later, scientists unearthed the tomb. They identified plants using impressions in the mud and phytoliths, microscopic shards of silicate in leaves. Plants from the mint and figwort families were especially abundant, species with aromatic flowers and leaves, and colorful sprays of blooms. In life, flowers connect the heavens and the underworld. Their blooms reach up into the aerial world, their roots burrow in soil. What better way to honor the transitions of death and burial than with these unifiers of realms?

Sages, including what we now call Judean and Greek sages, left distinctive marks in the graves: square stems and jigsaw-shaped phytoliths. Greek sage is a culinary and medicinal species, well known in the kitchen. I was not familiar with Judean sage, and so I bought some seeds and grew several plants in the plot that Katie and I tend in the community garden near our home in Atlanta. The first growing season produced only green stalks. The next year, the plants matured into armfuls of slightly fuzzy leaves and stems, topped with nodding clusters of purple blooms. The plants grew lush and waist high. I can see how ancient peoples would look to this plant for a funerary bed. The plant is soft and springy when laid down in bundles, with a much milder aroma than the sages we use for cooking.

The sage plants in the Mount Carmel graves are the first known use of colorful and aromatic flowers as funerary beds, inhumed along with dead people. These graves are among the first to have been organized in a cemetery-like way, in a distinctive location used over centuries. It seems that flowers and other aromatic plants may have been part of human funerary rights for as long as we've been burying people in long-term plots, at least in the Middle East.

The idea for fragrant grave linings may have come from the beds of vegetation that early humans slept on. In the Sibudu rock shelter in South Africa, remains of grass mattresses from seventy-seven thousand years ago are topped with aromatic Cape quince, a plant whose strong aroma repels insects. Humans are not alone in using scented plants to deter insects from their beds. Chimpanzees and other apes often use aromatic plants as they build their nightly sleeping platforms. Birds tuck into their nests strong-smelling herbs including, these days, tobacco-filled cigarette butts.

Cocooning the dead and protecting sleeping people are medicinal behaviors. Often the potency of plant medicine stems directly from the volatiles in its tissues. These chemicals kill pathogens and encourage healing. Nonhuman animals self-medicate with pungent leaves, and knowledge of how to use plants for healing is ubiquitous among human cultures. Chemical residues on the teeth of Neanderthals from forty-nine thousand years ago show that they were munching yarrow and chamomile, presumably for their medicinal properties. Until the advent of modern medicine, our ancestors used their noses to find and assess medicines. We do the same when we sniff a cup of tea. Teas with the best aroma have the most beneficial molecules. Smell guides us, and many self-medicating nonhuman animals, to the right plant and the right dose. Building on this biological foundation, cultures around the world use aromatic plants for medicine and funerary rituals. These cultural practices are also the foundation of modern perfumery technologies.

In ancient Egypt, aromatic plants were first used to make incenses along with the scented oils, tars, fats, and waxes needed for embalming. Mostly, the embalmers used tree resins. Techniques learned through the preparation of funerary aromas were later put to use extracting floral scents into unguents, oily salves, and water-based extracts. The development of perfumes in China and southern Asia show a similar pattern: Pungent plants were first used as medicines and in religious contexts, then aromatic oils and waters for secular purposes were developed, including those that use floral aromas.

In the first millennium BCE, carved relief on stone shows Egyptians pressing lilies for perfume. Assyrian clay tablets from about the same time record the medicinal use of rose water and the activities of royal perfumeries staffed by professional perfumers, often expert women. In the 330s BCE, Darius III, the king of Persia, employed fourteen perfumers and forty-six garland makers. Ancient Greeks and Romans scented not only their bodies, but wine, clothes, temples, and houses. Perfume, for them, was not just a matter of personal adornment, but a joyous collective experience. Xenophanes, writing about a Greek banquet in fifth century BCE, tells us in Richard Folkard's 1884 translation, "Each guest upon his forehead bears / A wreath'd flow'ry crown; from slender vase / A willing youth presents to each in turn / A sweet and costly perfume… / Fragrance of flowers, and honey newly made." Party time? Bring on the floral perfume.

When we perfume ourselves, we don't usually think about medicine and death rites. Usually, we seek the opposite, health and vitality. These are two sides of the same coin. Perfumes don't just cover up sickly body aromas, they literally cleanse the skin. Rose and jasmine oils, for example, kill *E. coli* bacteria, including strains that are resistant to antibiotics. Medicine and olfactory adornment have been linked since ancient times. The first-century CE Roman medical compendium and instruction book *De materia medica* by Dioscorides, for example, starts with detailed instructions for making scented oils from dozens of plants, including flowers of jasmine, rose, lily, narcissus, and henna. During plague outbreaks in seventeenth-century London, prices of flowers, garlands, and herbs surged, as good aromas were believed to be protective. The word *perfume* carries the memory of the medicinal effects of aroma in its etymology, from the Latin *per fumar,* "to come through the smoke." Only in the sixteenth century did the word lose its fumigant roots and come to mean a beauty product scented with flowers and other aromatics.

We perfume our skin, but the effects of plant aromas go deeper. They become part of us. Aromatic molecules not only bind to receptors in our noses, triggering signals to the brain, they also dissolve into our blood. From there, some of them pass through barriers that keep blood out of the brain, steeping our cerebral nerves. No wonder our blood pressure goes down, eyes dilate, body relaxes, cognition perks up, and memories become more solid after we smell a flower.

Like perfumer bees, we use plant aromas to change the chemical language of our own species. Human bodies make over a thousand different volatile aroma molecules. The scent of skin reveals our health, gender, emotional state, and individual identity. We're mostly not consciously aware of it, but these aromas affect how we feel and act. In the presence of the smell of anxious people, we tense up and make more risky decisions. We're attracted to the scent of our family members and, when we're looking for mates, to strangers whose aromas reveal them as genetically compatible. From body scent alone we can judge, with some accuracy, how neurotic, attractive, or socially dominant someone is. Or so it seems from a limited number of scientific experiments on human bodily aromas. Enter perfume, the beguiler and revealer.

At the simplest level, perfumes can mask our body's aromatic signals. By overwhelming my skin's aroma with rose oil or other perfumes, I clothe myself, blocking your nose's ability to directly sense my body. Perfumes have been shown in scientific experiments to blur people's perceptions of gender, assertiveness, and neuroticism.

Some studies, though, suggest that rather than masking our skin's aromas, we often choose perfumes that complement and highlight the individual scent of our bodies. This individuality is partly a result of the uniqueness of our immune systems. We can literally smell one another's immune system genes. A well-chosen perfume, then, is more like a beautifully tailored outfit than a veil. The wrong perfume is an ill-fitting pantomime costume. Trust your nose, not the marketing.

Mimicry is a last, intriguing-but-unproved possible cause of our love for floral aroma. A few of the aromas made by human skin are the same molecules as those made by some flowers. This overlap is partial. Unfortunately, we don't naturally smell of roses. But we do share a few volatiles with flowers. This happy accident—flowers did not evolve to mimic us, nor we them—means that when I smell a flower, a few of my nasal receptors tingle with human scent. From a scientific point of view, we have no idea what hidden messages might be conveyed by such scented illusion.

It is only in the last three decades that the genes for human aroma receptors have been discovered, a breakthrough that won scientists Richard Axel and Linda Buck a 2004 Nobel Prize in Physiology or Medicine. Today, we lack a map of exactly which aroma molecules stimulate which receptors. We are a long way from scientific understanding of how each aroma molecule interacts with others to stimulate the brain and change our thoughts, feelings, and behaviors. But what scientists don't yet know, perfumers have studied for centuries. When connoisseurs say that a perfume is lascivious, calming, or confident, they may be reporting how floral smells mimic human aromatic signals. Of course, this biological effect is vastly modified by cultural context, but underneath changing cultures, flowers and human skin may share some chemical language.

❦ ❦ ❦ ❦ ❦ ❦

Perfumes made of flowers evoke feelings of beauty and desire in others, with an undercurrent of illusion. This is exactly what flower aromas do in the wild with pollinators, and so our perfumes carry not only the material presence of flowers into the human realm, but some of the flowers' ways of being. There are no perfumes made of unmodified human scent, declaring to the nose, *This is who I am.* Instead, we present to the world a more complicated aromatic story, one that merges our identity with that of flowers and other more-than-human aromas. *This is who we are.*

Perfume tells the world that we are made from relationship, we are living communities. We transcend the "self," not as a metaphor or a mystical experience, but as a reality experienced in our bodies. There is a paradox here. To more fully express our humanity, we reach for plants. To be "me," human, I need "you," plant. This ability to create interbeing is part of the original genius of flowering plants, one reason for their wild success over more than one hundred million years. By creating experiences of beauty, flowers stitch creatures that were separate or only loosely connected into tight, intimate, and creative relationships.

When we give a scented flower, bring blooms to a grave, or dab perfume onto our skin, we are not enacting arbitrary, merely symbolic rituals. Rather, we invoke the relationships with flowering plants from which the ecology of the planet is made, and which created and sustain human life. No wonder we mark every major transition in our lives with flowers. A bouquet or perfume bottle invokes our creators.

Enjoyment of floral aroma is one way to connect with and understand how flowers shape our world. Another way is to use science. To find out how—and to glimpse the beauty and horrors that emerged from this work—we must leave the sunny gardens of the South of France and descend into a fortified basement under the streets of London. There, flowers that lived centuries ago lie preserved between sheets of paper. These blooms still shape the world today.

Tea

Flowers reveal hidden

histories

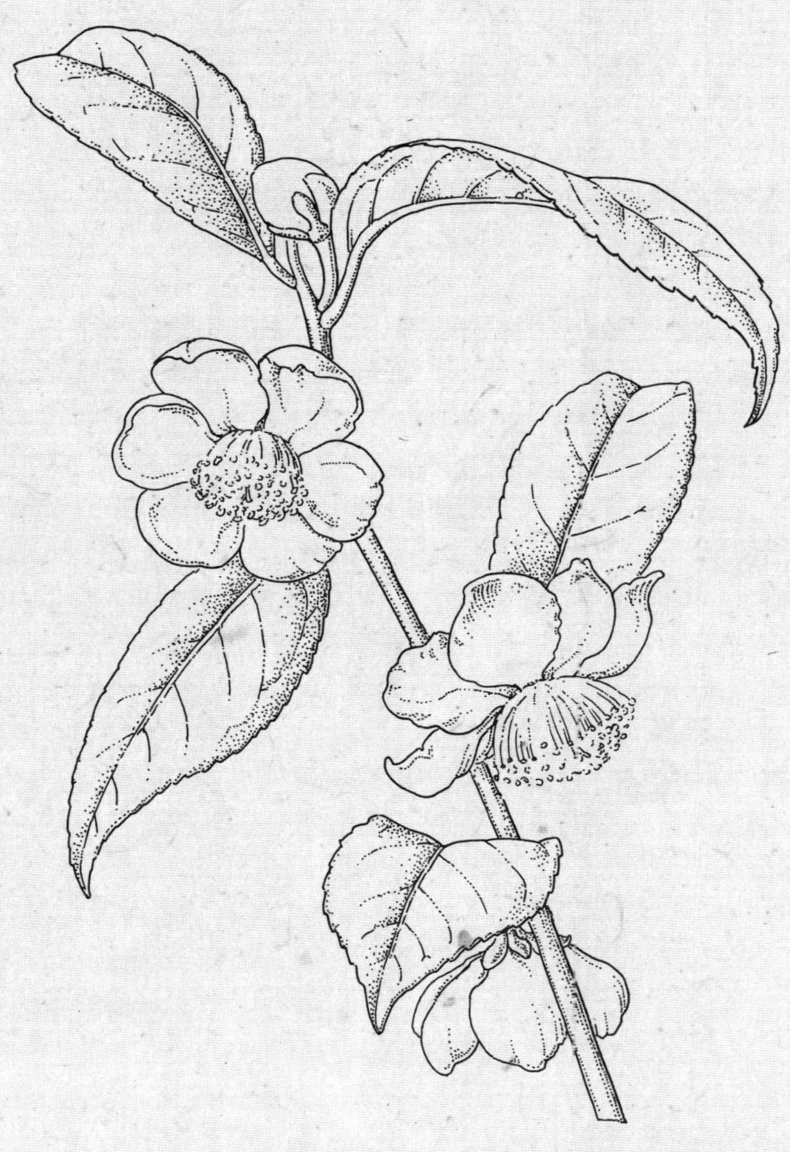

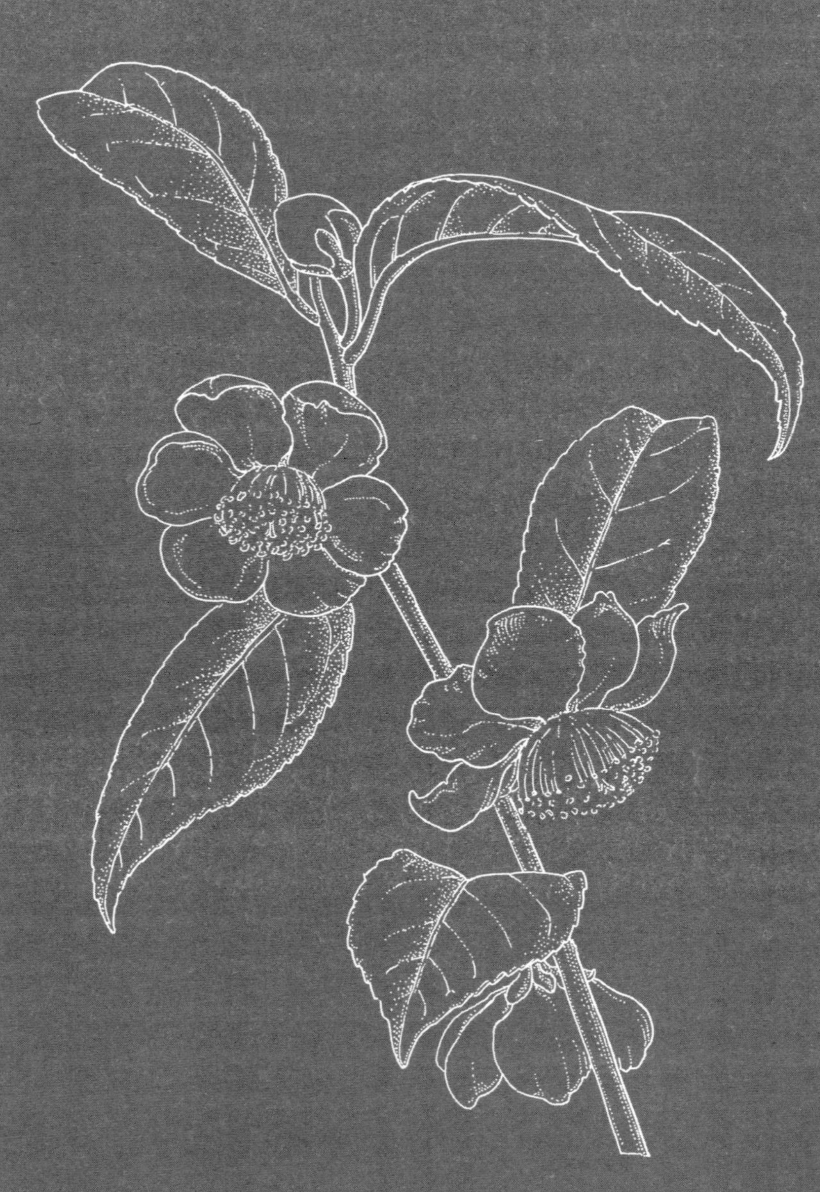

D R. ISABELLE CHARMANTIER, HEAD OF COLLECTIONS at the Linnean Society of London, unlocks a fortified door and we step into a bombproof vault under Piccadilly, the busy thoroughfare that runs through central London. The interior of the vault smells woody and warm, with a tang of something bitter, like old herbs. The walls are lined with bookshelves stuffed with old leather- and clothbound volumes, all atop wooden cabinets. This is the repository for much of the personal collection of books and dried plants of the Swedish naturalist Carl Linnaeus. The cabinets are filled with pressed flowers and leaves, the plants that Linnaeus picked, preserved, and studied during his lifetime. After the death of Linnaeus and his son, his impoverished family sold the collections in 1784 to a young Englishman, James Edward Smith, son of a wealthy wool merchant. Except for a brief protective excursion to the countryside during the blitz of World War II, most of Linnaeus's books and plants have lived in London ever since, the core of the collections of the Linnean Society, the world's oldest organization devoted to the science of natural history.

These dried flowers, and the books that describe them, transformed scientific understanding of the living world. Almost all modern biology is partly shaped by the insights that emerged from them. So, too, were some of the obscenities of colonialism. It all started with flowers and their sex lives.

"This is one of his first publications that uses the sexual system." Isabelle pulls a slim volume from the shelves and rests it on a leather-topped table in the center of the room. "Linnaeus published it in Leiden in 1735 when he was in Holland for a few years. He's twenty-eight when he publishes this, so extremely young." For a book that is nearly three hundred years old, it is in great condition, pages only slightly yellowed. Isabelle tells me that this book, only a dozen printed pages long, is the first edition of Carl Linnaeus's *Systema Naturae*, a founding document of modern biological classification. In it he elaborates on a plant classification scheme he first published in a pamphlet given to his landlord as a New Year's gift five years earlier. The book also categorizes minerals and animals. To this day, biologists organize nature's exuberant diversity into Linnaean categories like species, orders, and classes, groups that he delineates in this book. Few of the details of his system survive, such as which plant fits into which class, but the overall hierarchical architecture is almost unchanged. Over the next thirty years, Linnaeus expanded this work through ten editions and hundreds of related publications.

"Yes! This is it; 1737. Let's put it down here." Isabelle hefts a book the size of a jumbo accounting ledger onto the table. Opening this much larger volume, Isabelle reveals lists of Latin plant names in crisp, elegant type. Turning the oversized pages, she continues, "This is *Hortus Cliffortianus*, the catalog of the plants in George Clifford's collection at Hartecamp. Linnaeus also uses his sexual system in this book." In *Systema Naturae*, Linnaeus laid out his principles, taking just three pages to summarize his plant classification scheme. In *Hortus Cliffortianus*, he applied the scheme to a collection of living plants, organizing and showing off the botanical acquisitions of his wealthy patron in a five-hundred-page catalog.

The "sexual system" is Linnaeus's method of classification of plants. He grouped species with similar male floral sexual parts—the pollen-shedding anthers and their stalks—together, arranging the entire plant world into twenty-four classes. He divided each class into up to eleven orders, mostly using the arrangements of the female parts of the flower, the pollen-grabbing stigmas, their supports, and ovaries. In a few classes he added orders based on features of the male parts, the fruits, or whether the flowers were unisexual.

Linnaeus's floral method of biological bookkeeping reshaped biology, but not in the way that he intended. Like tricksters, the flowers showed him one truth, then revealed deeper, contradictory secrets to later generations of biologists.

The fact that the Linnaean sexual system made its first appearance while he was in Holland hints at another important way in which flowers shaped modernity. Linnaeus's employer, George Clifford, was a banker and one of the directors of the Dutch East India Company, a corporation that was one of the most wealthy, powerful, and cruelly destructive forces of colonialism in the eighteenth century. Today's billionaires build spaceships and buy media companies. In Linnaeus's day, they assembled plant collections and hired naturalists to tend, study, and classify them. This botanical largesse was not motivated by curiosity solely for its own sake. Colonial power was funded by the understanding and control of flowering plants. By seizing land and enslaving people, colonists gained the plants, land, and labor they needed to grow profitable flowering plants: cotton, sugarcane, tea, opium, indigo, coffee, cinchona, and dozens more. In colonial hands, "flower power" was a foundation for the horrors of empire. The creatures—flowers—that, in the past, had transformed the Earth largely through cooperation with others, were put to work in the most exploitative system humans had yet devised, one where extraction by the few replaced reciprocity among the many.

George Clifford was one of the richest businessmen in Europe, and his keen interest in collecting and displaying plants coincided with the source of his and his country's colonial wealth. The young Linnaeus was a hireling in that system, tasked with supervising the care of the plants in Clifford's gardens. Because Clifford was enmeshed in global trade and rich enough to build and heat four extravagant glasshouses, he grew tropical species from Africa and Asia, along with plants from southern Europe. Linnaeus, an obsessive botanizer from an early age, wrote that his first encounters with these plants on visiting Clifford's estate left him feeling "bewitched," "amazed," and "quite carried away." With the help of introductions from local patrons, Linnaeus met Clifford and surely impressed the wealthy plant collector not only with his enthusiasm and knowledge about plants, but with the power of his new classification system. By peering into a flower and examining its sexual parts, Linnaeus could immediately give a plant a place in one of his tidy categories, a classification that clarified the plants' identity and gave it an affinity with other plants. In modern terms, Linnaeus was a master of database management. For Clifford, a collector whose gardens were continually receiving new shipments of plants from colonized lands, this young enthusiast in need of a job was an attractive addition to the staff, organizing and tending to the botanical collection, and acting as a "physician" to advise on medicinal uses of plants.

In the vault, Isabelle points to the first entry in the opened pages of *Hortus Cliffortianus*, the giant catalog of Clifford's collection that places species in the sexual system and cross-references them to the published work of previous authors. "The first plant listed is *Canna* because it has one stamen [the male part] and one pistil [the female part]. And here's a specimen." She unties a ribbon around a package about the size of a large notebook, then unfolds the wrapping-paper sheets to reveal a plant pressed onto a sheet of sturdy paper. Although the plant's color has dried to tea-leaf brown, it is immediately recognizable as what gardeners today call a "canna lily," a colorful relative not of lilies but gingers. A large paddle-shaped leaf lies on the right-hand side of the paper sheet and a stalk rests along the center and left. One branch of the stalk has flattened spiky black balls attached halfway up, the dried remains of the plant's fruits. The other part of the stalk branches into a spray of about five flowers, their petals now a pale hazelnut brown. This was one of the plants that Linnaeus cut, then used to study and classify the species.

Unlike modern scientific herbarium specimens, Linnaeus's *Canna* plant sits frustratingly alone on the sheet, with no useful information about where or when Linnaeus collected and preserved the plant. Did it come from Clifford's garden or did Linnaeus collect it elsewhere? We cannot know. This flower's individual narrative is gone. In Linnaeus's hands, it has shed selfhood and history, and become a physical marker in an abstract scheme. The only annotation is an inked "1.1" in the top right corner, a cross reference to his classification likely added by nineteenth-century curators. No modern botanist would treat plants this way. The data and notes accompanying each specimen are now highly valued. I've been told by many curators that a plant sample (or fossil or preserved animal) without precise information about the date and place where it was collected is scientifically useless. It strikes me that in omitting this information we also might lose a sense of respect for the living being that we've taken out of the cycle of growth and decay to preserve in our collections. Unmarked graves swallow identities and stories.

There are fourteen thousand other preserved plants in the collection at the Linnean Society of London, and each fits into one of Linnaeus's classes and orders. Linnaeus counted a single stamen and pistil on *Canna* and so put the species in class Monandria, order Monogynia. (We now know that the flower is more complex than he realized, its "petals" being sterile male parts.) Flowers with two stamens, like olives and speedwell, he classified in Diandria, then Triandria for three-stamened flowers such as crocus and iris, and so on, with subgroupings for the number and arrangement of female parts. He grouped plants with more complex flowers not by counting, but by studying the sexual parts for their points of attachment, groupings, and fusions. Orchids, for example, he placed into the Gynandria for the union of their male and female parts into one structure.

Alongside the technical descriptions, Linnaeus includes notes that seem to take gleeful pleasure in the romantic and salacious potential of his scheme, freely drawing parallels with human sexuality. Among orchids, he tells us, "husbands are monstrously obsessed with women," referring to the physical fusion of male and female in the flower. In the Monandria, "a single husband in a marriage" describes the flower. But "two husbands in the same marriage" add a frisson of excitement to Diandria. The Polyandria get even more extreme, with "twenty or more husbands in the same room as a woman."

Isabelle unwraps another package from the herbarium collection, revealing blooms the color of clear mountain skies, the blue of *Delphinium* as fresh as if the flowers were cut yesterday. Linnaeus, as he often did, took great care to arrange the stems, leaves, and flowers in as natural positions as possible. The *Delphinium* seem to have flown straight from the meadow to the page. And there are the polyamorous parts, a mop of male anthers around a handful of female pistils. We now know that these blooms are not the "marriage beds" Linnaeus imagined. The pistils seek incoming pollen carried by insects from other flowers, and the anthers likewise send their pollen away. Unless the flower self-fertilizes—which usually happens only when insects fail to show up—blooms are less marriage beds than boardinghouses from which singles make forays into a dating scene elsewhere.

The overtly sexual nature of Linnaeus's method, helped by his heated annotations, gave other botanists pause. A dozen men in bed with one woman? Surely students might be "corrupted by the immorality that had broken out among the lilies and onions," wrote the prominent Russian botanist Johann Georg Siegesbeck in 1738, calling the system "loathsome harlotry." In France, the philosopher Julien Offray de La Mettrie produced in 1748 a satire, *L'homme-plante*, in which he lampooned Linnaeus's analogy of the botanical marital bed with mildly pornographic "botanical" descriptions of sexual anatomy and ejaculation in humans. The pearl-clutching continued for decades, with William Smellie, the Scottish editor of the first edition of the *Encyclopædia Britannica*, writing in 1773, that "obscenity is the very basis of the Linnaean system." As with other critics, Smellie may have used moral outrage to disguise professional jealousy. In 1765 he had published an inconsequential thesis about the sexes of plants. To see Linnaeus's ideas slowly being adopted, or at least debated, by many of his colleagues across Europe must have been galling.

At the time that he developed his system, Linnaeus was an impoverished young man from the rural backwaters of Sweden, with none of the wealth or social connections that helped so many of his contemporaries. He was a relentless self-promoter and so was surely aware that racy overtones were good for publicity. But titillation would only go so far. The lasting value of his system was its utility. Unlike the cumbersome classifications of his predecessors, Linnaeus offered a practical and easy-to-use method. Twenty-four classes are easy to remember, unlike the hundreds proposed by earlier botanists. His exclusive focus on flowers simplified categorization using relatively straightforward characters, consistently applied across all species. He even had a category, the Cryptogamia, for flowerless plants whose sexual parts were "hidden."

Previously, scientific botanists classified plants using hundreds of different characters, none of which were applied uniformly across all plants. Only people with extensive knowledge of plant anatomy could use these schemes. A keen student with a good eye could learn the Linnaean system in an afternoon. Little wonder that within decades of its publication the sexual system was used across Europe, albeit with strong opposition among a few holdout botanists. Even the Jardin du Roi in Paris, the gardens where French kings employed botanists to cultivate and study plants to learn their commercial and medicinal value, briefly experimented with organizing their plants according to the Linnaean sexual system. This despite the fact that many of the rival classification schemes had their origins in France.

Linnaeus's other innovation, giving every species a two-part name—*Quercus alba* for white oak and *Quercus rubra* for red oak, for example—also had the advantage of simplicity and uniformity. Others had used such double-barreled names before, but not with the same consistency across species and without embedding them in a larger classification of classes and orders. Previously, official names could run to whole sentences, with few governing rules about the naming process. Linnaean two-part names always start with the more general grouping, *Quercus* for all oaks, then the unique identifier for each species, *alba* or *rubra*. A database organized by simple and consistent principles is powerful and useful, unlike previous systems that he described as "vague, slippery, or variously applicable." In using two-part names, Linnaeus may also have tapped a tendency in human psychology to prefer doublet names. This naming convention is not universal—mononyms and three-or-more-parted names are used among some cultures—but in many places, people use two words to name and classify: forenames and surnames for people, make and model of cars, and general and specific names for the common names of species such as cranefly orchid, downy woodpecker, and oyster mushroom.

Linnaeus extended the practicality of his scheme to the everyday work of organizing collections of plants. Each flower and its leaves, pressed and dried between sheets of paper, served as an exemplar of a species, the reference against which others could be compared. Before Linnaeus, botanists mounted their specimens in giant bound volumes. Once a species was in position, it was fixed, and new pages could not be inserted around it. Linnaeus instead had carpenters build wooden cabinets with exactly twenty-four drawers. The sheets of paper that he used to mount his dried specimens were smaller than those used by most botanists. Each sheet slotted easily into his cabinets' openings. If new information emerged about a flower, he could refile the species. When new species arrived on his desk, sent to him from around the world as his prestige grew, he would slip each novelty into its place in the cabinets. Linnaeus's ideas and their physical manifestation in his cabinets were nimble, able to accommodate change and discovery.

Although it was a useful database management tool, Linnaeus's system was also a way of claiming ultimate control of naming. He sequestered to himself, and those who would pick up his work after his death, the sole authority for giving each species its official, scientific name. He preferred Latin and Greek for names of species and forbade in his system names given by the people who knew the plants best, the local human inhabitants of each plant's home. Let locals use their vernacular, he wrote, but "all learned Botanists should agree over the Latin names [or else] I foresee barbarism knocking at our gates." To this day, scientists use Linnaeus's 1753 *Species Plantarum*—an extension of *Systema Naturae* that lists individual plant species names—as the starting point for scientific names for plants. By using the Linnaean system, we value database consistency over local knowledge. We turn living kin into abstract categories. In the biblical book of Genesis, Adam's first God-given task was naming species. By this account, the "oldest profession" is not sex work, as Rudyard Kipling and others have claimed, but the naming of species. Linnaeus took the task of naming upon himself, with the explicit intention of centralizing authority. As we shall see, this approach fit with his twin project of bringing under Swedish economic control plants from around the world.

Linnaeus's system also embodies a creationist view of life. All species were, he claimed, "produced by the Infinite Being in the beginning." He thought he was giving names to species created in their present, unchanging forms. The idea that his neatly classified species might split, morph, or otherwise evolve was anathema. Yet, paradoxically, embedded within Linnaeus's classification is strong evidence for evolution. He failed to see an unexpected truth blazing from his system.

A treelike diagram occupies the entirety of the sixth page of the first edition of *Systema Naturae*. The tree is rooted on the left of the page, a trunk called Nuptiae Plantarum, "Plant Nuptials." As it grows across the page, the trunk splits repeatedly into branches. Each split is defined by a floral character such as "husbands and wives enjoy one and the same room" versus "husbands and wives enjoy separate rooms," referring to whether flowers are bisexual or unisexual. By the time the tree reaches the right side of the page, it has divided into twenty-four twigs, each one a Linnaean class of plant. Linnaeus intended the tree as a handy summary. But what emerges is an imprint, albeit an imperfect one, of the branching patterns of genealogy.

Flowers don't look similar for arbitrary reasons; the similarities among their sexual parts often reflect their ancestry. All orchids have fused male and female parts not because each independently evolved this feature, but because they inherited their anatomy from a common ancestor. Sometimes, similarity is due to independent invention in distantly related species. Iris, bluebell, chicory, indigo, and some columbines are blue not because they are close kin, but because each found that blue was a great way to attract local pollinators. But if we look at enough characters, kinship outweighs the happenstance of convergent evolution, and our classification becomes a sketch of ancestry.

Linnaeus had drawn an imperfect evolutionary tree but he didn't realize it. The flowers were giving him a glimpse of their kinship, but he regarded his classification scheme as no more than a useful tool. How wrong he was. By building a classification scheme based on flowers, Linnaeus revealed to others what he himself did not believe: that life evolves.

This evidence was one of the foundations of the Darwinian revolution. In *On the Origin of Species*, Charles Darwin wrote that even if the theory of evolution was "unsupported by other facts or arguments," the patterns revealed by classification would cause him "without hesitation" to conclude "that the innumerable species, genera, and families of organic beings, with which this world is peopled, have all descended, each within its own class or group, from common parents." Darwin wrote that classifications are not just about "mere resemblance," but reveal "something more," namely kinship and history. He called Linnaeus, along with the French zoologist Georges Cuvier, who extended Linnaean classifications, his "two gods."

Once Darwin and others pointed it out, biologists never again looked at classifications in the same way. Every classification was a time machine, allowing us to peer back and see treelike kinship relationships among species. Today, biologists routinely use the insights of these kinship or evolutionary trees to study the world. We design flu vaccines based on which branches of the virus tree seem most troublesome. We answer questions about when humans evolved using evolutionary trees. Likewise, we use the branching patterns of these trees to ask questions such as "Are fungi more closely related to plants or to animals?" "Which wild grasses gave us modern wheat?" or "Are these two plants the same species?" Flowers, through Linnaean classifications, were among the first creatures to teach us how to do this.

Linnaeus's flower-based classification drew on a very small number of characters, the arrangements of a handful of sexual parts in the flower. With such a limited range of data, his tree was an often misleading sketch of plant ancestry, one that was superseded in the decades after his death. Today, biologists use not only flowers, but all the features of living beings, including thousands of useful characters from DNA sequences, to reconstruct the past branching patterns of life. The primacy Linnaeus assigned to the arrangement of male parts in the flower was especially problematic. The male parts of a flower are but a few features among many and they lead to some strange bedfellows in the Linnaean scheme, such as the grouping of *Canna* with mare's tail, a feathery plant we now know is completely unrelated to the ginger-like *Canna*. Nonetheless, despite the imperfections of his scheme, it was flowers that first showed us how shared features of living species could illuminate patterns of kinship. Why? What is so special about flowers that they revealed some of life's history and evolutionary process?

Flowers were good teachers partly because they evolved to speak to animals. Their colors, aromas, and sizes evolved for the purpose of advertising themselves. DNA might be wonderfully informative about genealogy, but it is invisible without advanced technologies. Flowers' long evolutionary history of drawing attention had the unexpected effect of revealing to one clever species of ape—humans—the fact of evolution itself. From nectar to knowledge. That's quite the impromptu turn.

The sexuality of flowers also made them great revealers of kinship. When two plant species split from one another, dividing one branch of the family tree into two, flowers are often the mediators of the divide. A shift in flower aroma, color, or shape is enough to fracture one species into two, as we saw with changes in the aromas of orchids. When Linnaeus picked flowers for his classification scheme, he unwittingly chose the plant parts most likely to reveal the branching pattern of history.

Flowers' role as sexual mediators also means that they evolve at a moderate pace, making them excellent identifiers for individual species. When two new species emerge from a single ancestral species, their flowers usually evolve some distinguishing features but don't completely reinvent themselves. Two new orchids still look like orchids, each with its own quirks. New goatsbeard species still look like spiky dandelions, but the flower of each new species has its own petal length and color pattern. Flowers, then, have just the right balance of conservatism and innovation to make them excellent for species-level classification. If Linnaeus had used other, less changeable characters such as the shapes of leaves or the presence of woody stems, his classification would have been too coarse to be useful. If he had used highly variable characteristics, such as the number of leaves on a plant, he could not have reliably classified individual species. Flowers are Goldilocks characters, changing at just the right pace to identify species and reveal ancestry.

Linnaeus's system also, I think, succeeded because it fits with our everyday experience of flowers. We humans are a face-reading species, and flowers are very much like little faces. Just as our eyes linger on the faces of animals, including other humans, we look to flowers to understand the identities and personalities of plants. One needs no botanical training to see that the bowl-like flower of an apple blossom is like a little rose, or that lilies are like drooping tulips, or that most orchid blooms share a mannerist vibe. In these impressions we're glimpsing kinship. The pleasure we get from viewing flowers enlivens and makes memorable these insights.

Through Linnaeus's work, flowers spoke to us, revealing stories and processes that others had missed. He was a strange conduit for this plant-borne wisdom. He was vain and avaricious, consumed by professional jealousies, and often cruel to those around him. Yet, through this miasma of unpleasantness, he somehow formed a fruitful bond with flowers. Perhaps his mind and emotions were so wrapped up in plants that the social norms of dealings with humans eluded him. He would not be the first socially awkward "plant person" living at human social margins, as the lives of green-thumbed shamans, plant healers, "witches," and forest-dwelling mystics sometimes attest.

In addition, the fact that Linnaeus came from humble origins and was mostly self-taught gave him insights unavailable to others. Had he had the "advantage" of a proper education, he would have been immersed in the overly complex schemes of contemporary botanists. Instead, he learned directly from the plants. As a young man wearing worn and ill-fitting clothes, his shoes so ragged they were plugged with paper scraps, he wandered for hours in gardens and fields, then pored over botanic manuals in libraries, sketching and taking notes, seeking patterns among the blooms. For all his many failings, his manic dedication to flowers taught us something important: that the profusion of species on Earth is not a jumble. Instead, beautiful order lives within the diversity, an order that reveals ever-branching patterns of kinship. In the end, classification teaches us that we're all family.

❦ ❦ ❦ ❦ ❦

"The interesting thing about Linnaeus is that we look back at him as a scientist, doing work for pure curiosity and love of the natural world." Isabelle Charmantier lifts another set of paper sheets onto the table in the vault. "But it's not. A lot of it is economic." She opens the old paper sheets and reveals *Camellia* plants, dried to the color and texture of crusty old seaweed. Each has a browned bloom at the top of its branch. "Linnaeus thought Sweden was spending far too much money on luxury goods, and tea was one of those." The dried specimens from his collection are the few material remains of his lifelong quest to smuggle live tea plants—tea is made from one species of *Camellia*—from China to Sweden. The project aimed to set up Swedish tea plantations and undercut imports from China. At the root of this work was Linnaeus's love of money and his anxiety about what he thought were the financial losses of international trade.

Linnaeus often wrote of "barrels of gold" as a measure of an individual's and a nation's worth. Later in his life, when he'd become wealthy, he loved to pull out his gold coins, feel their heft and texture, and display them to penniless students, a real-life Silas Marner with a needy audience. This attention to gold permeated his work. As he named and described plants and animals, he kept a sharp eye on the economic potential of each species. When the Swedish king ennobled him in 1761, giving him the name Carl von Linné, it was not for classification or nomenclature, but for his work applying Chinese methods of pearl seeding to Swedish freshwater mussels. This pearl-farming experiment came to nothing, but at the time promised to revive depleted local supplies of pearls and enrich Sweden's coffers. Linnaeus commemorated his elevation to the aristocracy with a coat of arms crowned with a flower, the twinflower, a creeping form of honeysuckle common in subarctic habitats all around the Northern Hemisphere. The twinflower was his favorite bloom, one whose leaves he promoted as a "Lapp tea," an alternative to imports, even though his own son described the resulting brew as "rather repulsive."

Although his pawing of gold seems crass, Linnaeus's obsession with generating and hoarding wealth was rooted in experiences of extreme privation. As a young man, he often went hungry and could afford no good clothes, a miserable experience anywhere, but especially during Swedish winters. His parents would have told him about the great famines of the 1690s, the decade before his birth. Linnaeus regularly saw emaciated beggars in the university hamlet of Uppsala in the 1740s through the 1770s. He also lived through localized famines when poor citizens, in his words, "wither away, but also, which is even more gruesome, must listen to their little children's whimpering, suffering and death agonies." He had a keen sense of what happens when the flourishing and fruitfulness of plants is inadequate to meet human needs. It especially pained him to see chest loads of silver, the currency of choice for foreign trade in those days, leaving Sweden to make the sea voyage to buy tea and other goods from China. All these experiences lit in him a passion for discovering and promulgating new uses for Swedish plants. He also sought to bring to Sweden plants from around the world, not just to describe them in his classification, but to grow them locally after acclimating them to Sweden's winters. This, he thought, would stem the hemorrhage of precious metals and create homegrown prosperity.

His recommendations for backup foods during famine ranged from plausible to laughably impractical. Wild leeks and tree resin might help the starving. But tulip bulbs "cooked with butter and pepper" or black currants combined with "powdered sugar" suggest a disconnection from the reality of famine. Confident to the point of callousness, Linnaeus stated that "he who knows his plants shall never need perish during the crop failures," an assertion that presumes that his fellow Swedes, including those who worked year-round in the fields, were so ignorant of plants that they starved surrounded by plenty. None of his schemes for alternative foods caught on. Instead, more stable production of rye, turnips, and other crops warded off famine.

Linnaeus idolized national self-sufficiency, and he had a deep aversion to trade. If only Sweden could free itself from dependence on others, the barrels of gold would surely accumulate. This isolationism contrasted strongly with the market economics that Adam Smith and others were developing at the time. For them, foreign trade was the culmination of a nation's development. Economic exchange, not hoarding, ultimately increased prosperity as each nation traded the goods it could most efficiently produce and imported others. There is a loose parallel here with the ascendency of flowering plants. They owe their success to partnership and exchange with other species, not to walling themselves off and going it alone. Flowers and humans alike flourish in relationship with others and wither in isolation. For all his attention to flowers, Linnaeus failed to grasp this lesson.

Contact with people and lands overseas was of interest to Linnaeus primarily because it yielded new material for his classification and schemes of national self-sufficiency. This was not trade, but appropriation. As his reputation grew, Linnaeus attracted students, whom he called "apostles" or "disciples," and sent them around the world in search of plants, especially plants of potential economic value like mulberry for silk or tea for home cultivation. These explorers were ill-prepared and poorly funded, and many died during their peregrinations. Those who did return received little renumeration or recognition, although some survivors found later success by publishing work under their own names or by working with other botanical explorers such as the English botanist Joseph Banks.

Linnaeus's quest to bring to Sweden plants from around the world for cultivation was partly fed by an early floral triumph in his life. As a young man in Holland, he and Clifford's gardener, Dietrich Nietzel, coaxed a banana plant into bloom, a feat no one in Holland had yet managed. The banana plant, shipped in from an unspecified part of the Americas, grew in one of Clifford's heated greenhouses. In Linnaeus's words, for "five whole years . . . it showed no signs of flowering; but, lo! At the very beginning of this year 1736 . . . it presented its much longed for flowers," some of which matured into fruit. Crowds came to see the marvel. Linnaeus marked the achievement by sending a banana to his rivals in Paris, where bananas had, so far, failed to flower. He also rushed into print a fifty-page illustrated book, *Musa Cliffortiana* (Clifford's banana). The book celebrates the flower and its fruit; explores the anatomy, sexuality, and uses of bananas; and further promotes Linnaeus's sexual method of classification.

In the vault at the Linnean Society of London, Isabelle shows me a folder hand-inked in old, curly script with *Musa*, the scientific name of the banana. The folder contains paper sheets bearing three dried banana flowers. Each one is a stubby dried sheath with anthers barely protruding from their ends. Whether these came from the Clifford plant, we cannot say, but they mark a turning point in Linnaeus's thinking. If bananas could bloom and fruit in Holland, what other plants might be coaxed into flower away from their homes? The banana had bloomed thanks to heat from a greenhouse, but Linnaeus believed that plants could also toughen up outside and acclimate to colder climes. He vastly overestimated the physiological adaptability of plants. Although he tried for decades, none of his subtropical and tropical imports survived for more than a few years.

Bananas were a rare novelty. Tea, though, was a mainstay of international trade and a special obsession of Linnaeus. With no supporting evidence, Linnaeus declared in 1741 to the Swedish Academy of Sciences that "there is no doubt any more" that tea would grow in southern Sweden. "Poor Chinese," he gloated, they will "lose through this more than 100 barrels of gold a year." In Scandinavia's most widely read publication of the time, the farmer's almanac, he looked forward to Swedish tea plantations causing China to lose its "flowering bliss, and leaving Sweden not a small part of its fortune and happiness." His plans were repeatedly thwarted. Five of his apostles tried and failed. Then, two "tea" plants that arrived via the Swedish East India Company turned out to be different species of *Camellia*, not the kinds from which we make tea. The seeds he cajoled from botanists in London failed to grow. Live plants withered at sea or were eaten by ships' rats, making Linnaeus "inconsolable." In 1763, two live plants arrived in Uppsala, prompting Linnaeus to write to the ship's captain, "Truly if it is tea, I shall make your name, Mister Captain, more eternal than Alexander the Great."

In 1765, the long-hoped-for tea plants were growing outside, and the buds opened to reveal fragrant cups of white petals, with a crowd of yellow anthers. Finally, tea was blooming in Sweden. The celebration was short-lived. After the autumn bloom, the Swedish winter froze the plants to death.

Linnaeus's overconfidence in his ability to get plants to grow in Sweden condemned his attempts to establish local plantations of valuable plants. But the way in which he linked the study of flowers and the pursuit of economic gain paralleled the work of other botanists in Europe. The era of global trade was partly ushered in by the study of flowers and flowering plants.

Consider another caffeinated plant beloved by eighteenth-century Europeans, coffee. In 1714, French traders brought live coffee plants to the Jardin du Roi, the king's center for botanical study in Paris. At that time, processed coffee beans had been imported to France for decades, but no one in France had studied or propagated the living plants. At the

Jardin du Roi, Antoine de Jussieu, professor of botany, made an extensive anatomical study of the living plants, including flowers and fruit, after he succeeded in inducing the plants to bloom in a heated greenhouse. He also germinated coffee seeds. His 1715 treatise on the plants reads like a dry botanical description, but this seemingly obscure academic text directly seeded French colonialism. Unlike Linnaeus, Jussieu did not try to grow coffee commercially in Europe. Instead, he propagated the plants and dispatched the resulting young saplings to French colonies. In this way, French plantations in the Indian Ocean and western Atlantic became lucrative hubs of coffee production.

In London and Holland, too, botanical gardens served as research centers and nodes in trade networks for plants that would ultimately be grown on colonized lands, usually tended by enslaved or indentured people on plantations. Linnaeus had many plants trafficked to Sweden for his research and failed plantations, but he had no direct role in moving plants out of their places of origin to colonies. Linnaeus did, however, provide intellectual justification for racist methods of production, including enslavement.

In *Systema Naturae*, the publication in which he introduced his sexual system of floral classification, Linnaeus also discussed animals, including humans. He was the first Western naturalist to publish a classification that included humans listed in the same category as other animals, in the subgrouping that also included monkeys and apes. Other naturalists protested, yet Linnaeus insisted that humans were animals and that all humans belonged to the same species. But Linnaeus also divided humans into varieties: "European white, American reddish, Asian tawny, African black." By the tenth edition of *Systema Naturae*, in 1758, Linnaeus expanded on his human classification, adding behavioral and moral descriptions, asserting without any evidence that whites were "sanguine, wise" and blacks "lazy, neglectful." Although he did not use the term *race*, this edition of *Systema Naturae* wrote racist beliefs into the database that almost all Western scientists used as their guide to biological classi-fication. Because Linnaeus was at that time the most well-known and influential scientific authority on the classification of life, his white-supremacist claims had enormous weight, as did his belief that humans could be divided into categories with unifying behavioral features.

Linnaeus was not directly an enslaver or colonial plantation owner, but his racism provided others with "scientific" justification for enslave-ment and other forms of racially stratified exploitation. What started as a system to classify flowers developed, through Linnaeus's own elabora-tion of his classification into the human realm, into part of the intellectual foundation for the monstrosities of empire founded on enslavement. In many ways, Linnaeus was a planter in every meaning of the word. Shortly after Linnaeus's death, King Gustave III of Sweden acquired the Caribbean island of Saint Barthélemy from the French and used favor-able tax laws to turn it into a regional hub for the trade of enslaved people. That Sweden's—and the Western world's—leading scientific authority on the classification of life wrote white supremacy into his opus surely was one enabling factor.

Flowers have histories hidden within them. Some of these histories are about their kinship and evolution. The architecture of flowers con-tains the imprint of genealogy, and so careful attention to flowers is a way of looking back into time. But such a gaze is never objective, nor are its elaborations. Linnaeus did not transcribe the truths of nature. Rather, he glimpsed a few and spun the rest out of conjecture and prejudice. He built racism into the foundation of modern biology. Others followed. Charles Darwin, while he was an ardent abolitionist, nonetheless be-lieved in the superiority of whites over "degraded" "savages" and wrote in 1871, in *The Descent of Man*, that "the civilised races of man will almost certainly exterminate and replace throughout the world the savage races." In the early twentieth century, many biologists, including lead-ing evolutionary biologists, were eugenicists, and their beliefs in "planned breeding" and "racial improvement" paved the way for coerced mass sterilization campaigns and were a foundation for Nazi genocide.

One lesson of Linnaeus's flowers and the classification that they inspired is to give us a stark reminder that science is always wrapped in the cultural context of its time and the prejudices of its practitioners. Linnaeus shoehorned his preconceived ideas about human diversity into a system that only seemed objective. In our time, racism propped up by erroneous and distorted "science" is on the rise again, and "science" is invoked to impose rigid dualities of gender and racial classifications within regressive political movements. The Linnaean legacy is a reminder that such prejudice is especially pernicious when it appears to have the authority of science, allowing falsehood to masquerade as rigor, a deadly illusion. The risk is especially acute when scientific authority is controlled by a small, homogenous community such as the eugenicists of the early twentieth century, or a single individual such as Carl Linnaeus, "Prince of Botanists."

❦ ❦ ❦ ❦ ❦ ❦

It is not just in a vault of centuries-old dried flowers or the foundations of biological science that we find floral illusions. The everyday blooms in our gardens, flower shops, and public green spaces also carry damaging secrets of a different kind.

When humans and plants intertwine their lives through gardening and the horticultural trade, extraordinary reciprocity and creativity emerges. But by shaping and controlling flowers, we sometimes also cause wounds, often unanticipated, that hide behind a mask of superficial beauty. These are hidden floral histories of our own time. Ones we can rectify. To find them, let's visit some prizewinning gardens.

Pansy

Beauty and brokenness in the garden

Look like th' innocent flower
but be the serpent under't.
—Lady Macbeth

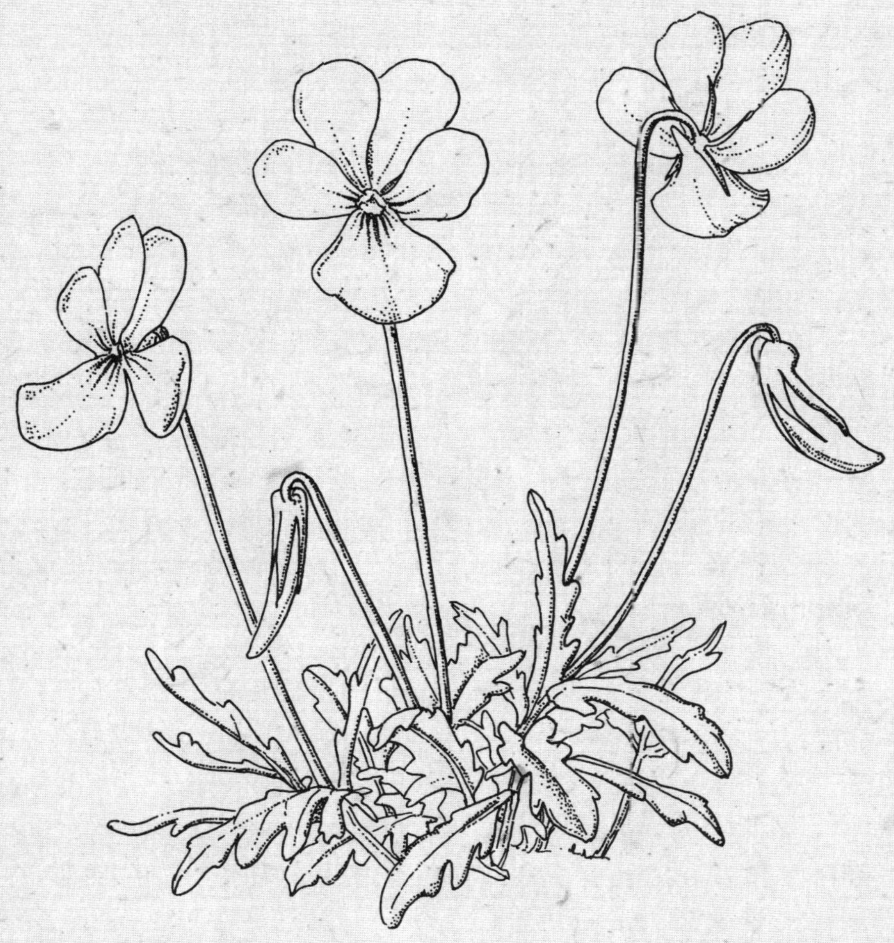

FIRST PRIZE, PANSIES AND VIOLAS," "FIRST PRIZE, Pansies." In 1950 and 1952, my great-grandfather swept the pansy competition at the Christchurch Horticultural Society, a gardening club on the south coast of England. He also won awards for rhubarb and marrows, a form of large summer squash, but growing perfect pansies was his special talent. The cards naming the prizes are tucked in a file box between his First World War army shoulder patch, some tattered black-and-white family photographs, and a few of his watercolor sketches. The box lives at my sister's house in Scotland, a small physical connection to the third generation before us. In the 1950s, pansies' association with gay and gender nonconforming men was well established—flourishing especially in the American "pansy craze" of drag performances in the 1920s and 1930s—but nothing in my great-grandfather's remaining belongings tells us whether this symbolism was significant for him. We just know that he loved pansy flowers and could coax perfect blooms from his garden.

As we unpack the box and look at the photographs, I feel a flash of recognition. My great-grandfather died a decade before I was born, but these photographs might as well be from my camera roll, minus the gardening awards. Here is Reg Cantelo standing in his small patch of garden, vegetables sprawling around him, flowers growing up to his waist. Here he is hugging a magnificently fluffy cat. Man and cat squint skeptically at the lens. The generations pass, but our dispositions echo. I write with a cat sprawled on my desk, his head lolling over the keyboard, and I look through a window at a garden of veggies and flowers. Lenses directed at my face make me uncomfortable. Later today, I'll tend our plot in the community vegetable garden. My homegrown pansies, though, will earn no prizes. They burned up weeks ago under the Atlanta sun.

A World War II food rations book sits alongside the other mementos in the box. Although vegetables were not rationed, a precipitous wartime drop in imports caused shortages of all veg and fruit across the United Kingdom. For Reg's generation, gardening was a "Dig for Victory" patriotic duty for anyone with the privilege of a patch of ground to tend. From 1939 to 1942, the number of home vegetable gardens in the UK nearly doubled to 5 million, and garden allotments increased from 960,000 to 1.7 million. Reg's garden was a small one, tucked behind a semidetached house, yet alongside his vegetables, he devoted space for flowers. My late grandmother, his daughter, told me of Reg's pride at both feeding the family and growing lovely blooms. I'm just guessing, but after serving in one war and enduring the bombing blitz of the south coast during a second, by the 1950s he and Dorrie, my great-grandmother, may have been hungry not only for food, but to have their spirits lifted by the pleasures of homegrown flowers. Even if we are lucky enough not to live in a modern war zone, we can relate to this feeling. In a world beset by fractures and tumult, flowers in gardens, window boxes, and bouquets offer sensory portals into the vitality of the plant world, reminders that, despite it all, life can be beautiful.

The award cards don't tell us which pansy varieties won Reg his prizes, but he had plenty of choice. By the 1950s, hundreds of domesticated varieties were available as inexpensive seeds. The names of early twentieth-century pansy varieties speak of the possibilities: Deep Blue, Bronze, Fancy Red, Purple Mantle, Marbled Dark, Large Spotted White, Yellow Blue Edge, and Giant Trimardeau in eleven different colors. What riches! Reg was an amateur painter and was surely drawn to the varieties with unusual or spectacular colors. With his mind on the horticultural show, some of the giants might also have appealed.

The profusion of colors and shapes available to Reg had recent origins. Unlike roses and peonies that have been selectively bred and hybridized by people in Asia and Europe for millennia, the wild Eurasian pansy, *Viola tricolor*, was not cultivated until the early nineteenth century when a frenzy of crossbreeding by horticulturalists produced hundreds of new varieties. Seed catalogs reveal the change. Kennedy and Lee's 1774 *Catalogue of Plants and Seeds* published in London lists for sale eleven species of European and North American wild violet, but no pansies, not even wild ones. Goring & Wright's 1798 plant catalog, also from London, lists no pansies and just one "double blue scented violet." But by 1835, botanists J. Sinclair and J. Freeman published a booklet devoted to the "nearly one thousand varieties of Panseys," plants that had been produced by "several Ladies of rank and fashion" who had elevated the pansy above its former stature as a "noxious weed." In 1899, Dicksons' seed catalog, from Chester, England, tells us, "Of late years a very great advancement has been made in the improvement of Pansies. The flowers are now larger & finer, & of greater substance, whilst the variety in colour is much more extensive." The seeds were also carried across the Atlantic, where they and the imported wild pansy are known as Johnny-jump-up for their rapid growth and self-seeding. In 1888, Philadelphia-based Burpee Seeds sold over one hundred thousand packets of pansy seed, calling them "the most popular of all flowers grown from seed."

The new varieties not only expanded wild pansy's color palette into pastels and reds, they also often featured dark face-like central markings and frilled petals. Most were two or three times the size of their wild ancestors, although a few were bred into miniatures. In 1868, the biologist Charles Darwin noted with admiration the "great contrast between the small, dull, elongated, irregular flowers of the wild pansy, and the beautiful, flat, symmetrical, circular, velvet-like flowers, more than two inches in diameter, magnificently and variously coloured, which are exhibited at our shows."

Darwin, I beg to differ. Wild pansies are anything but "dull." People in western Europe have been celebrating their beauty for centuries. Their jaunty petals and flat faces give them a friendly, humanoid look. They naturally come in all shades of purple, white, and yellow, splashing color along European field and wood margins. Herbalists use the flower, leaves, and stringy roots to clear the chest and to treat pain. Folk names abound, most of them referring to love, presumably because the flowers seem so joyful: heartsease, come-and-cuddle-me, jack-jump-up-and-kiss-me, love-in-idleness, and tickle-my-fancy. The name "pansy" comes from the French, *pensée*, both for lovers' dreamy thoughts and scholars' dedication to the mind. King Louis XV of France honored the studious physician and economist François Quesnay with a three-pansied coat of arms. Shakespeare mentions heartsease as a love potion and pansies for thoughts, and a young Queen Elizabeth I embroidered heartsease onto a prayer book cover for her stepmother. All these references were to uncultivated wild pansies.

But such wild beauty was not enough for nineteenth-century English horticulturalists. Using the natural variability of the flowers as a starting point for crossbreeding experiments, they reshaped the wild pansy. The work in the 1810s of Mary Elizabeth Bennet, along with her gardener, William Richard, was especially influential. Bennet was the daughter of aristocrats who gained their wealth and leisure partly through the forced labor of enslaved people in North America toiling not in gardens, but in the brutal conditions of plantation fields. Here, as with Linnaeus and Jussieu, human injustice and exploitation were entangled with the study and manipulation of plants. In their English gardens, Bennet and Richard collected and interbred as many colors and forms as they could find. Their fancy pansies soon attracted the interest of commercial seedsman James Lee, who then worked with Bennet and Richard, and other aristocratic plant enthusiasts, to interbreed their novelties with imported varieties. Within a decade, plant breeders hybridized the Bennet-Richard plants with other wild pansy species from European mountains and eastern Europe, adding yet more variety. Most garden pansies from the mid-nineteenth century to the modern day are cross-species hybrids.

In some ways, the work of these nineteenth-century breeders continued a much older natural process. There are over one hundred wild species of pansy around the world, each with its own flower color and shape. Many of these wild pansies first evolved by natural hybridization of what were separate species. More broadly, studies of the genetics of the nearly six hundred species of *Viola*, the genus that encompasses all violets and pansies, show that more than two thirds of these species evolved by hybridization. The violet and pansy family pedigree is full of interweaving, interlocking strands. Human hands elaborated lacework already underway.

From the point of view of pollinating insects, the new pansy varieties were duds and a severe break from the past. From the insects' perspective, the pansy was no longer what it seemed, a rich source of nectar and pollen. The new varieties looked good to humans, but were mostly useless for insects. Whichever variety Reg chose, the flowers impoverished the ecology of his garden compared with wild, "weedy" pansies.

When Mary Elizabeth Bennet and her collaborators sought to enhance wild pansy, they unwittingly severed the plants' connections to pollinators. The flowers changed almost overnight, and the insects were left behind. Today, in side-by-side tests, wild pansies are generally better for pollinators than big or frilly cultivated varieties. The artificially bred cultivars sometimes have long nectar-filled spurs, but few insects can access this bounty. Short-tongued bees have no hope of reaching the nectar and even long-tongued bumblebees struggle, unable to gain a grip on the huge petals.

Wild pansies were built by evolution to appeal to the aesthetic preferences of insects. The shape and size of their petals, their petal colors and guidelines, and their aromas match the body shapes and senses of pollinators. When human aesthetics intervene, we prioritize features that appeal to us, not to pollinators. We're huge creatures compared with insects; a human eyeball is twenty times the size of a honeybee. So we often value exaggerated flower sizes and big color splashes rather than the smaller, finer sizes and colors of many wildflowers. Frills, extra petals, and other novelties catch our eyes. We don't intend to, but our work can undermine the very relationships with pollinators that first built the flowers. Now, to get pansy seeds, commercial seed growers often use needles or tiny brushes to hand-pollinate flowers, producing hybrid seeds of known parentage. For some varieties, self-pollination of flowers also maintains hybrids.

Pansies are just one of many hobbled flowers. Visit a modern garden center and, although flowers abound, few serve insects. In some flowers, an inherited mutation causes the flowers to grow extra petals, a process called "doubling," although the flowers often have far more than twice the typical number of petals. The May roses in Grasse, for example, are descendants of doubled roses mentioned by the Roman writer Pliny two thousand years ago. In China, doubled peonies were cultivated by 750 CE. Today, pom-pom-like daisies are doubled, as are columbines and larkspurs that look like Elizabethan ruffs. Petunias, soapworts, and bird's-foot trefoil all come in doubled variants. In many, the fuss of extra petals blocks access by pollinators to nectar, and in some, the nectar-producing parts of the plant are gone. Human aesthetics are once again at cross-purposes with the needs of insects. Doubled flowers often grow their extra petals by converting what would have been pollen-making stamens into petals, a mutation that starves insects of protein-rich pollen. Not all doubled flowers are hopeless for insects, though. As I saw in Grasse, some doubled roses offer pollen, as do some marigolds and a few others. But most doubled flowers offer insects the illusion of a generous bloom, but little or no food. *Come feast here*, scream the shopfronts, but the shelves are empty.

The gulf between what we humans like in pansies and what pollinators need is so wide that, for insects, cultivated pansies are like aliens arrived from a different world, "planet human aesthetics." For other garden plants, the arrival from other worlds is literal. When hummingbird-pollinated American columbines, lobelias, and salvias find themselves in Europe, where no hummingbirds live, the flowers please human eyes but offer little food for local animals. The reproduction of the plants, too, suffers. Transcontinental transplants must either self-fertilize, never fertilize, or make do with the inept attentions of local pollinators. For most, propagation by humans is the only way to breed. Such loveless transplants are common in ornamental gardens. South American tobacco in our garden grows white tubular flowers, each as thin as a pencil and as long as my forefinger. In their Andean homes, moths unfurl prodigiously long proboscises to pollinate the flowers, but when the flowers perfume the evening air of Atlanta with their rich, musky scent, no local moth can reach the nectar.

Not all global plant travelers are so bereft. Those shaped like cups, short tubes, or wide funnels are accessible to animals wherever they grow. Even far outside the native ranges of the plants, beetles eat magnolia flower pollen, butterflies sip nectar from zinnia blooms, and honeybees feed on dandelion and clover. There is a limit to the hospitality of these transplants. They may provide some nectar and pollen, but only to pollinators with broad tastes and flexible behaviors. Others are left out, such as specialized pollinators like hummingbirds and bumblebees, whose beaks and proboscises are adapted to feed on local flowers, and solitary bees that feed exclusively on the pollen of a few flower species.

The translocated plants also offer little in the way of food for caterpillars and other young insects. This is a disaster for insect life. Animals like nesting songbirds that rely on caterpillars as food also suffer. Insects evolved, often over hundreds of thousands of years, the ability to detoxify and digest the leaves of plants native to their homes. Exotic transplants often arrive with leaves defended in ways that few local insects can deal with. The lack of chewed holes, scars left by leaf miners, and other signs of insect life on these imports might please humans seeking visual perfection, but such plants are wastelands for insects. Evolution has not had enough time to adapt insects to the plants.

A study of the hedgerows of the mid-Atlantic region of the United States, for example, found that those dominated by non-native flowering plants like autumn olive, bittersweet, honeysuckle, and multiflora rose have depleted insect communities compared with hedgerows thick with local black cherry, white oak, viburnum, and sweet gum. Sixty-eight fewer caterpillar species and 91 percent fewer caterpillars live in the non-native hedges. Little wonder that insect populations are crashing, and that nesting birds are struggling. Their food supply has collapsed.

To human senses, novel plant ecosystems look green and full of blooms, but this is an illusion of ecological vitality. Without tens of thousands of years of coevolution to adapt to the flowers and leaf chemistry of the plants, animals cannot thrive. There are exceptions. Fruiting plants like honeysuckles often feed birds, at least for part of the year. Imported plants that belong to the same genus as locals tend to have similar chemistries, and so local insects find them easier to feed on. But even among these plant imports with close local kin, insect abundance is often two thirds lower than on the locals.

Our fascination with importing attractive plants from elsewhere has sometimes upset local ecologies far beyond gardens. Worldwide, 60 or more percent of the hundreds of imported plant species that smother or otherwise disrupt local ecosystems first arrived in their new homes thanks to the horticultural trade, including knotweed, lantana, callery pear, and privet. This international trade is still underway. Plant breeders scour the world for wild species to cultivate and sell in the ornamental plant industry. Given that translocated plants are wiping out some local ecosystems and causing billions of dollars of damage to agriculture, this ongoing quest to find and move ornamental plants around the world seems improvident, at best.

We should be careful about building xenophobic walls through our gardens—native plants good, immigrant plants always bad—and instead think at a community level. Does this garden or landscape sustain the diversity and productivity of local species, as well as meet human needs? Growing a small number of plants that please the human eye but do little for local ecologies seems a joyous and benign expression of our desire to live close to beautiful flowers. Our fascination with the shapes, colors, and aromas of plants from around the world is a sign of our loving curiosity about the plant world. For many of us, nonlocal flowers hold personal value. In my garden, the Japanese irises remind me of the close friends who dug and shared them with me when we moved to Atlanta. As an immigrant, I am drawn to some European flowers, especially springtime bulbs, that remind me of my childhood home. Others—our rose bush and daisies—are there for sensory delight, gifts that my partner Katie and I have given each other. Most garden plants do not jump the fence and take over the surrounding landscape. Especially in higher latitudes, we lift our spirits by supplementing the relatively small number of local species with cheerful additions from other places. But if plants from elsewhere dominate, to the exclusion of locals, we have impoverished the life of the garden. Our aesthetic experience, too, suffers. We lose the capacity or incentive to seek and cultivate the many beauties of home rather than always searching for stimulation from elsewhere.

Gardens built around local species feed our human need to be close to thrumming hubs of life. When I'm in the doldrums or feeling beaten down, watching dozens of bees and hoverflies scrabbling around native sunflowers, asters, or goldenrods restores me, as does the sight of birds plucking caterpillars to ferry to squalling nestlings. I don't quite know how this renewal works. Intellectually, perhaps it is about hope. But the experience feels less like a reorientation of the mind and more like a wordless reconnection of my body with the sustaining sensory richness of life.

The work of reweaving local ecologies, garden by garden, adds up to a potentially vast change in the ecological health of the land. In many cities, industrial areas, and agricultural landscapes, gardens are among the few places not paved over or turned to monoculture crops. For example, one quarter of London's land area is made up of 3.8 million private gardens. In the United States, 55 percent of people engage in some form of gardening, and lawns occupy 1.9 percent of the continental terrestrial area, three times the area of any irrigated crop. Paradoxically, "rewilding" our world depends on the most "domestic" of activities: tending our window boxes, allotments, gardens, and verges with enthusiastic attention to local ecology. The "wild" exists not only in remote hallowed protected areas where humans are largely excluded, but also where we live and work. To plant a vibrant garden is to hallow our homes.

You can sense the difference. A garden full of native plants hums with insects. Birds dart and swoop, animating the undergrowth and sky, and even the soil smells richer, deeper. Try it at home. I planted horticultural marigolds and chrysanthemums next to local goldenrods and asters, arranging them side by side outside our back door in Atlanta. Two golden plant species and two purples. What a difference in insect life! On warm days, dozens of bumblebees and a handful of skipper butterflies, wasps, and small bees crawled over the local plants from dawn to dusk. Many slept through the night nestled in the blooms. The horticultural flowers were eerily empty. Even though they were inches from hubs of insect life, the only visitors they received were occasional bumblebees that landed, sat for a moment, then winged off. On the goldenrods and asters, the ancient relationships between form and function of flowers were alive and well. The marigolds and mums offered only a "still life" to human eyes, a ghastly glimpse of a world where insects are gone or irrelevant. The wild ancestors of these plants fed insects—in

Central America for marigolds and China for chrysanthemums—but in our Atlanta garden these nectarless cultivars and dislocated imports might as well have been made of plastic.

Plastic flowers might be preferable to the doses of pesticides that many commercially grown ornamental plants bring to our gardens. Nowhere are the destructive shadows of horticulture more troubling and damaging than in our use of chemicals. I never experienced Reg's garden shed, but I vividly remember those of my grandparents, all of whom were avid flower and vegetable gardeners. The tang of creosote planks and garden chemicals greeted our noses as we entered the small, rickety shelters. Neatly lined up on the shelves, above the jars of old screws, were rows of cans and bottles, each one containing a different poison. On the labels were names and pictures of the intended victims. Slugs and snails. Caterpillars. Fungi. Mildew. Beetles. Weeds. My grandparents loved birds and other animals, and so applied chemicals with caution. Nonetheless, the point of their arsenals was to knock back unwanted animals, plants, and fungi.

The concentrations of poison used today in gardens and ornamental landscapes can be staggering. Ornamental plants and trees are often treated with soil drenches, chemicals added in liquid form to the base of the plant. These drenches cause roots to draw the chemicals into the whole plant. In the pursuit of beauty, we turn entire plants poisonous. Nursery-grown potted plants can have pesticides applied at 250 times the level used to treat corn before planting in agricultural fields. Ornamental trees can get well over 1,000 times this level. These numbers assume that whoever is applying the poison is following industry standards. Amateurs often slosh the pesticides, herbicides, and fungicides at even higher rates.

Walk into a conventional garden center and there's a good chance that many, perhaps most, of the plants there contain toxic synthetic chemicals. A 2017 study of flowers for sale at garden centers in England, for example, found that nearly all contained insecticides and fungicides, not just in the leaves, but also in pollen, which is the main food of young bees. A few plants were suffused with ten different added toxins. The ornamental plants in the study were all of species known to be attractive to bees and caterpillars. The combined effects on insects of what the authors of the study called a "cocktail of pesticides" are not fully known, although many of the chemicals, by design, disrupt the behavior, digestion, and reproduction of insects, and some were present in the samples at concentrations high enough to cause harm.

Milkweed plants—the sole food of the monarch butterfly—bought at nurseries throughout the United States are often contaminated with poisons. A 2022 study found 61 different pesticides in samples of 235 milkweed plants from 33 retail nurseries. The average plant contained over 12 different pesticide residues. For the handful of these chemicals whose effects have been studied, concentrations were high enough in the nursery milkweed plants to harm the behavior and growth of monarch butterfly caterpillars in at least 38 percent of leaf samples in the study. Appallingly, plants labeled as "wildlife friendly" were almost twice as likely to contain at least one pesticide at a level known to be harmful. A 2024 study in Sweden of garden center and supermarket plants labeled as "bee friendly" found twenty-one substances known to be harmful to bees, "at worrying levels," along with dozens of other chemicals of unknown effect. Some lavender flowers in garden centers are suffused with carcinogenic fungicide, and other flowers contain substances known to be carcinogens, mutagens, or toxic to human reproduction. That's not the bouquet or basket of garden flowers we want to give or get. Studies of chemicals in garden shrubs and trees show that single applications of horticultural toxins can leave residues for a year or more, enough to sicken or kill any insects that nibble on their leaves or eat pollen or nectar.

Things are just as dire in much of the cut-flower industry, where flowers are often grown on one continent, then flown for sale on another. A study of roses, gerberas, and chrysanthemums from Belgian florists' shops and supermarkets found 107 different poisons on the flowers. Roses were the worst, averaging about 14 chemicals per bloom. A global review of the cut-flower trade found over 200 pesticides and fungicides used on the flowers. Unlike rules that apply to edible plants like vegetables, fruits, and herbs, importing countries set no maximum limit on levels of horticultural chemicals.

"Loathsome canker lives in sweetest bud." Shakespeare's thirty-fifth sonnet reads now like an augury for modern flowers. Bouquets are illusions, beautiful to the eye but suffused with poison. Although many of the flowers in the horticulture trade look wonderful and sometimes even smell good, this seeming wholesomeness is built on a brew of chemicals specifically engineered to destroy life. Can such flowers be beautiful?

In the garden, the destructive chemicals sicken and kill insects and other flower- and leaf-feeding animals. In the greenhouses and fields that supply the ornamental gardening and cut-flower markets, it is also sometimes humans and surrounding communities and habitats that suffer. Agricultural chemicals can cause neurological, reproductive, and lung problems in workers. Growers often wear protective equipment, although enforcement of safety standards is sometimes lax. Residues in dust and water runoff contaminate the land and water around fields. This off-site contamination affects humans as well as wildlife. A study of children in Ecuador found that the closer the children's homes were to flower-growing facilities, the worse were their "neurobehavioral" symptoms such as lack of attention, poor language skills, and difficulties with memory and learning. "Insofar as these roses were beautiful," wrote Rebecca Solnit in *Orwell's Roses* after visiting a Colombian greenhouse, "their beauty was meant to occur somewhere else, for someone else, a continent away." A rose jet-propelled from South to North America is the epitome of the aesthetic and therefore moral dissociation that happens with global trade. With no sensory connection to the consequences of our actions, our ethics are unrooted.

The problems continue when cut flowers arrive at their destinations. Florists seldom wear protective equipment even though the flowers they work with can have concentrations of toxic chemicals sometimes one thousand times higher than allowed on food plants. Consumers, too, are potentially exposed. A study of potted garden plants and cut flowers for sale in Germany and Austria found that more than 40 percent of potted plants and 72 percent of cut flowers contained pesticide residues harmful to human health. Some of the chemicals detected were not approved for use in the European Union, and so the chemicals were either residues leftover in soil from previous years, used illegally, or imported from countries with laxer regulations.

Not all growers use prodigious quantities of poisons. A rare few shun all chemicals. Others apply chemicals only when problems emerge, instead of applying them prophylactically to all plants as is standard practice in much of the industry. If we consumers want plants with minimal or no chemical applications, the onus is on us to discern whether "wildlife friendly" labels mean what they say. But chemical testing is hard to find and, when I checked in 2024, costs more than one hundred dollars for each leaf sample. Finding local growers with an ecological approach is the best bet.

Governmental regulation has been slow and often weak. In 2018, the European Union banned the use of neonicotinoids, a class of pesticides that are especially harmful to bees. The ban applies only to field-grown crops, not greenhouse operations or imports. The ban also caused many growers to switch to other pesticides as they sought to control pests in a cost-effective way. In the US, neonicotinoids and other pesticides are routinely used on field crops and in horticulture, with minimal regulations for labeling of the resulting products and application guidelines designed to protect commercial honeybee hives but not wild pollinators. If you're looking for help to protect your garden from chemicals, or even seeking reliable labeling information to guide your shopping—is the "bee friendly" label just greenwashing?—governmental agencies have done little to help.

Despite these woes, I think it is too easy to point a sanctimonious finger at people in the mainstream horticultural industry. The nectarless, ecologically displaced, and poisoned flowers in our gardens, supermarkets, and florists' shops grow from a web of interrelated causes. No part of the web is singly responsible, and tweaking individual strands is unlikely to change the whole. We have been selectively breeding flowers to meet human desires, but not the needs of pollinators, for millennia. We consumers demand perfect-looking blooms. For imported flowers, customs agents comb through blooms in search of potentially problematic imported insects, and so insecticides ease the passage of flowers across borders. Pesticide manufacturers and governments play regulatory cat and mouse, with different outcomes in different countries, and huge financial incentives for chemical companies to delay and minimize restrictions. In regions where horticulture is a major industry, hundreds of thousands of jobs depend on the trade. In a market with little regulation or labeling, and very small profit margins, the financial incentives for growers to chemically treat plants are intense. Aphids, whiteflies, mites, thrips, scale insects, leaf miners, caterpillars, gnats, and fungal disease all delight in greenhouses or fields packed with tens of thousands of often genetically uniform plants. It is from this web of interlinked causes that the plants of the modern horticultural trade emerge.

Like many others, Reg Cantelo, my pansy-growing great-grandfather, was unwittingly caught in a web where floral beauty is not always as wholesome as it seems. His cultivated pansies fed far fewer insects than

England's gorgeous native pansies. The chemicals that he used probably also eroded the vitality of his garden. In our pursuit of beauty and fruitfulness, we inadvertently cause harm, often to the very life processes that we want to nurture in our gardens. Today, the illusion is even more perverse and invisible than in Reg's time. When the majority of blooms for sale at the garden center have little relation to local ecologies and are tainted with poison, we're in a world of dreadful make-believe. Unlike our ancestors, for whom flowers were unambiguous signs of good habitats, we're faced with a world where the beauty of a bloom often cannot be trusted.

This illusion causes a tragedy of misdirection. Gardens can teach us what it means to live in fruitful relationship with flowering plants. There, we participate in an ancient reciprocal bond between flowers and animals, started by the first flowers more than one hundred million years ago. Gardeners already know this in our bones. When we garden, we tap into some ineffable but vitally important and powerful source of renewal and meaning. Reg felt it, I'm sure. I feel it, alongside much frustration when inevitable failures beset my plants. Even the frustration feels grounded. Yet, modern horticulture too often diverts gardening into ecologically damaging or sterile ends.

I feel trapped. I adore the brightness and gorgeously varied forms of flowers at the garden center and florist's shop. The flower display is often the best part of the supermarket. Few things lift the spirits more than an amble through a flower-filled public garden. Even mannerist pansies are a joy to see. Yet these pleasures have a bitter heart. It need not be this way.

Flowers should give life, not destroy it. An absurdly simple idea. To make it real, we need to reimagine our relationships with flowers. Today, we mostly think of flowers as ephemeral and powerless, as disposable ornaments. Instead, let's see flowers for what they are: ancient life-givers whose beauty builds networks of cooperation. Beauty, in this view, is not just present in superficial appearance, but in the processes that built and sustain flowers. This sounds abstract and idealistic, but the cultural change is starting, a nascent revolution of imagination with practical consequences for how we grow and manage flowers.

❦ ❦ ❦ ❦ ❦

The Chelsea Flower Show in London is like the Academy Awards in the United States, doling out prestigious awards with much hoopla. Even if the event is of no personal interest, you know it's happening. Prime-time television news programs cover the show, and royalty and celebrities make sure to be seen there. Winning garden designs are featured in newspapers and online magazines. Across the UK, the Chelsea Flower Show is a symbol of posh and polished gardening, a pinnacle of contemporary horticulture, and a bellwether for gardening trends. The show also serves as a foil for more modest gardening attempts. Like many others in the UK, my family members use self-deprecating sarcasm to comment on their struggling blooms: *Well, that's not going to win the Chelsea Flower Show, is it?*

Run by the Royal Horticultural Society (RHS)—a charity with over six hundred thousand members, more than the British Labour and Conservative Parties combined—the show is a flagship and fundraiser for the rest of the organization. The RHS maintains public gardens, offers formal and informal education, conducts horticultural research, advocates for gardens in local communities, advises government, and offers gardening advice to beginners and professionals through manuals, short books, special reports, and electronic media. Founded in 1804, the RHS is one of the most trusted organizations in the UK. They offer beautiful gardens and sound advice, projecting competence and a wholesome love of gardening.

I visited the show in 2023, and my daylong experience swung from wonder, to alienation, then to cautious optimism. These swings were partly caused by the vastness of the show and the range of gardening philosophies crowded together. Hundreds of garden designers and vendors each bring their ideas and wares, from no-holds-barred garden construction, to displays of the very best of single flower varieties, to gardens built to honor causes, from mental health awareness to wildlife conservation. The show has an upper-crust vibe: Champagne stalls and "luxury fine dining" tents intermix with vendors offering expensive garden equipment and buildings.

Plant wonder is easy to find at the Chelsea Flower Show. The vast central pavilion is an emporium of perfect-looking flowers, like a gathering of Platonic ideals. Tidy domes of chrysanthemums, every petal unmarred and each bloom exactly symmetrical, display hues as distinctive and rich as oil paints. I don't usually find chrysanthemums particularly appealing, but I was awed by this altar to their perfection. Clematis and sweet peas offered blooms so large and variously colored they seemed like tropical butterflies swarming the vines. A large metal carousel was densely hung with fuchsias. I had no idea that these flowers could be so lush or that they came in such subtle gradations of reds and pinks. Gladioli spikes were nearly as thick and tall as human legs. Delphiniums grew as tall as me, their colors a distillation of every shade of sky. A display of orchids—each one at the very peak of its bloom—demonstrated the diversity of pollination strategies, a rare nod to the world of flowers in the wild. Roses were arranged in the pattern of a formal garden, the blooms thick on each shrub. Another part of the pavilion hosted cut-flower arrangements sumptuous enough for the marriage of gods: bouquets two or three meters tall and wooden garden archways aflame with blooms.

Overlooking the displays was a singing Eurasian robin atop a central statue in the pavilion, a bird at once at home and out of place. No doubt this was his perch before canopies covered what are, for the rest of the year, open lawns and walkways on the grounds of Royal Hospital Chelsea, a home for veterans located on the banks of the Thames River since the seventeenth century. The robin sang for most of the day, his rolling, chattering notes a beautiful and disconcerting reminder that this temporary gathering of floral beauty is embedded in a world beyond, a world where real animals are seeking to make their homes amid our creations.

Outside the pavilion, the wonders continue. The thirty-six show gardens are arranged on two sides of the pavilion, with smaller garden displays along the other sides. The world's top designers compete to build these show gardens. Winning a medal at the show is a great honor, and the event also allows designers to showcase their ideas and, for some, a charitable cause. The largest show gardens are ten by twenty meters, about the size of a tennis court, the smallest eight by six meters. These dimensions roughly match the median size of a garden in the UK.

The show gardens look grown in place, with a permanent feel, but this is an illusion. Every plant, stone, pond, or garden structure is hauled to the site, planted, dug in, or built during the first two weeks of May. The gardens are open to the public for less than a week, then disassembled and unrooted by the end of the month. Like theater, the show requires a suspension of disbelief. Each show garden is on display for about as long as a bouquet of flowers lasts in a vase. But the skill of the designers, horticulturalists, and builders makes the gardens feel not only rooted, but timeworn. Trees look as if they've been in place for decades. Poppies and wild grasses bloom at the edges of boulders as if in a natural meadow. Ponds and streams seem part of the contours of the landscape, with mosses creeping alongside them. Yet, if we take a few steps along the pathway that connects the show gardens, we are wrenched from one staged reality into another. In a ten-minute amble, I was immersed in the sights of mountain woodlands, Mediterranean villas, British meadows, brownfield industrial sites, and placeless formal gardens.

Walking among the show gardens is strangely emotionally and intellectually demanding. Each garden has its own aesthetic and narrative.

Unlike the blasts of color and form inside the pavilion, the show gardens offer more complex stories. Like reading a book, it takes a little while to adjust to the voices of the authors and to "get" each story. In the morning, when crowds were thin, this reading was a pleasure. By midafternoon, the show gardens were thronged. Getting to most was like squeezing through the scrums at a packed rock-and-roll festival, with floral dresses and pressed linen substituting for tie-dyes and cutoff shorts. I was glad to have had less harried time earlier in the day.

A few show gardens were shockingly grandiose. Using brick, paving, and sometimes hundreds of tons of quarried rock, they built entire mountain brooks and large portions of buildings, adding mature trees, manicured reflecting pools, and tens of thousands of plants. The perfection of these creations spoke of power and wealth. Their feel was aristocratic, like English and French monarchs building and shaping "nature" for their pleasure. The flowers and other plants in them felt, to me, dwarfed by the scale of human ambition and achievement. But installed in permanent locations and cared for over decades, I could imagine how each of these designs could, in fact, restore beauty and even ecological integrity to sites that were perhaps mere lawns before. Context matters, and the Chelsea Flower Show is a place to explore and show off what is possible, not to turn that possibility into ongoing reality. In the end, though, this horticulture for the fabulously rich, impressive though it is, hardly offers realistic options for most amateur gardeners or professional designers.

I found more delight and hope for the future in the many show gardens that foregrounded indigenous plants and included "sustainability" as a core design principle. Wild City Studio, led by designers Jon Davies and Steve Williams, collaborated with the Centre for Mental Health to create *The Balance*, a garden set in a postindustrial urban setting, one that integrates human and nonhuman well-being, and that aims to be affordable. Built using recycled crushed building-site waste for paving, seating, and some of the matrix for rooting plants, the garden was lush with flowers and small trees, all centered around a rusted metal grotto, a reclaimed steel-clad shipping container in which edible mushrooms grew. Many of the plants were "weeds" like herb Robert, wild garlic, and bramble, interplanted with species such as bladder campion that can thrive in challenging places. Many were native to Britain, mixed with edible plants from elsewhere: North American blueberry to feed humans, southern European purple toadflax to satisfy nectar-hungry insects, along with edible herbs and small fruiting trees.

Even in the thick of the afternoon crowds, *The Balance* garden felt welcoming and calm. Seeing so many healthy plants bursting from broken concrete and rough gravel was uplifting. I felt not so much the "balance" of the garden's name, but an eruption of life as flowering plants restore places we have broken. This restoration is not redemption—flowering plants do not erase the traumas we cause the world and one another—but their vigor and adaptability do offer defiant renewal amid brokenness. Am I reading too much into some artfully planted flowers in a simulacrum of a neglected urban area? Perhaps, but flowers built our world, which means they can also rebuild. Flower shows, for all their artifice, directly connect our senses to this green potential.

Other show gardens also celebrated native plants and ecological design. The garden of the Royal Entomological Society, planned by designer Tom Massey, emphasized that gardens give us places to meet insects in the everyday. Wonder, curiosity, and respect can grow from these encounters. Most of us live in structures and landscapes where insects are absent or merely annoying pests. But in a garden, we see and hear bees, hoverflies, ants, caterpillars, and myriad other insects going about their lives. It is a symptom of how estranged we've become from the community of life that such connection to insects should be something we need to design for. But unless we're lucky enough to live near a nature reserve or an agricultural landscape that has not been turned to monoculture, where else but in a garden or planted park can we experience the relationships among insects and plants that built modern ecosystems? Massey's design encouraged the insects and our connections with them using native plants like poppies, viper's bugloss, hazel, and dandelion. Nestled within the garden was an "insect eye" seven-meter dome made from hexagonal tiles. Inside the structure, microscopes and information guides invited curiosity and exploration. The lower walls of the dome were made from stacked timber and sticks, excellent nesting sites for insects, as were the rammed-earth walkways and deadwood throughout the garden.

These and many of the other show gardens were relocated to new sites when the show closed. *The Balance* will become part of new community gardens in Tottenham's Markfield Park in northeast London. The Royal Entomological Society's garden moved to a mixed-use residential area near Queen Elizabeth Olympic Park in east London. This reuse is part of the sustainability strategy of the Chelsea Flower Show and the designers who bring their work there. Formerly, design of show gardens focused only on the week of the show. Each show garden was a very expensive single-use item, discarded after a week, sending hundreds of tons of concrete, plastic, and plants to the landfill. The parade of dumpsters, mostly not on public view after the show's closing, was the long shadow of the show's floral beauty. This throwaway culture not only wasted energy, construction materials, and plants, it kept the wonders of the show sequestered behind high walls and entrance fees. Now some show gardens and their plants have afterlives that reach into the community.

Pollinators and native plants were stars at the show not only in many gardens, but also in the booths of charities and community groups. The virtues of peat-free gardening were another repeated theme. Harvested from ancient bogs, peat has been the main ingredient of most planting mixes since the mid-twentieth century, favored for its lack of soilborne diseases, good water-holding capacity, easy drainage, and low cost. But peat mining not only destroys long-term stores of carbon, it wipes out wetland ecosystems. The Chelsea Flower Show and all the RHS's gardens aim to become peat-free, albeit on a timeline that keeps getting delayed, and many vendors and displays at the show featured peat alternatives.

Pesticide use, though, got no mention at the show that I could find, except indirectly in the few gardens that emphasized their organic growing practices. The unplanned but sometimes dire ecological consequences of plant breeding were also unspoken. Yet pesticides and selective breeding are foundations for the extraordinary gathering of floral beauty on display here.

That not a word is breathed at the Chelsea Flower Show about the damaging chemical and genetic foundations of modern horticulture seemed to me, at first, a horrifying omission. More, the silence felt indefensible at a time when the biological diversity of the planet is in precipitous decline, a crisis especially severe in Britain, where populations of many flowering plants, insects, and birds have plummeted over the last century. If we can hear at the show about peat and pollinator-friendly flowers, we can also hear about the soil drenches, sprays, dusts, and pellets that undergird much of the industry. The sometimes unfortunate consequences of flower breeding, likewise, would surely resonate with the curiosity of flower-loving visitors and might open a few imaginations.

But perhaps the Royal Horticultural Society is being more strategic than my knee-jerk disbelief and disappointment. Hasty anger or judgment are easy to find. Actual solutions to complex problems like the culture, economics, and ecology of gardening and flower production require a subtler approach, especially for a well-regarded charity like the RHS that can persuade and advise but has no power to compel change. At a flower show, the cause of planting native flowers is a positive message, a "yes" to plants and their insect collaborators. Peat-free compost, too, is going mainstream, no longer controversial although many commercial growers are still hesitant about making the switch. Banning or curtailing nasty chemicals is a more abrasive "no," one that butts directly against everyday practices in horticulture. Questioning the results of plant breeding perhaps even more so. The flowers at the show were largely bred to please us, not to serve pollinators or other flower-dependent species. The Chelsea Flower Show is not so much a celebration of flowers as a showcase of what people have done to flowers. Part of the suspension of disbelief at the show involves keeping this power asymmetry—human agency reigning supreme—unseen in the background.

Yet, that very fact that we humans are so cleverly devoted to shaping the floral world is, potentially, a foundation of a future horticulture that more explicitly includes the needs of other species. The more ecologically minded show gardens are excellent examples. If we are powerful and knowledgeable enough to re-create flower-filled woodlands and Mediterranean villa gardens for a one-week show in London, we also have the power to build gardens that ignite local biodiversity. If we can breed giant, frilly pansies or pom-pom-like chrysanthemums, we can also grow beautiful plants that invigorate habitats rather than starve or poison them.

The Chelsea Flower Show is celebratory. Perhaps there isn't room at such a festive occasion for critical examination of our gardens and the horticultural industry that supplies them? On the other hand, the show could be an ideal place for thoughtful engagement with the shadows and hidden brokenness of our pursuit of floral beauty. Flower lovers don't need shielding or diverting. We are not so naive to believe that a garden is an escape from every travail. The opposite is true. Gardening puts us in intimate touch with the earthy, green troubles of the living world, from failed plantings, uncooperative weather, difficult soils, and the balance of control and acceptance that every "pest" and "weed" challenges us with. Bound up with these difficulties are experiences of beauty, fruitfulness, and the pleasures of labor rewarded. A garden is a place to live into those tensions, heartbreaks, and joys. That plant breeding and chemicals add a few more twists to the plot hardly breaks the overall narrative.

Can we both celebrate floral beauty and ask better of ourselves? In their advocacy and educational materials, the RHS answers yes. The Chelsea show raises funds for the organization's outreach and publications, the go-to sources for gardening wisdom in the UK. Two linked messages now guide this RHS outreach. First, gardens are good for us. Second, gardens need to be good for the world beyond the human. These generalities find expression in granular plans and down-to-earth advice. How to manage a hedge to trap air pollution. How to encourage native flowers in a small garden. How to use plants to slow water runoff from your garden. Which plants create the most pleasing soundscapes? If insects are troubling your flowers, how do you deal with them without reaching for the spray bottle? The RHS also funds new community gardens and seeks to make its gardening advice accessible across all communities, to "open up gardening for anyone, anywhere." One in eight Britons currently has no access to private or shared gardens, and this privation is much higher among minority and lower-income communities. If well-being grows in gardens, then unequal access is an injustice, one that the RHS is focused on addressing.

Beautiful flowers are worth celebrating for their own sakes. Floral beauty also, when fully seen, shows us what needs mending in our world. For all its studious avoidance of friction, the Chelsea Flower Show is an unsettling party. I left feeling the elation that comes from immersion in over-the-top floral perfection and beauty. I was also encouraged to see among some displays an embrace of local plants, fungi, and animals, along with commitments to empower people to garden in places traditionally neglected by planners, designers, and horticulturalists. When dandelions and building rubble are highlighted at the Chelsea Flower Show, as they were in a couple of show gardens, we know that a cultural shift is underway.

✼ ✼ ✼ ✼ ✼ ✼

In the months immediately following my visit to the Chelsea Flower Show, some of Britain's most prominent gardening TV and social media personalities decried native plantings as misguided and harmful to cherished horticultural traditions. The Best Show Garden award from the previous year went to *A Rewilding Britain Landscape*, designed by Lulu Urquhart and Adam Hunt. The garden was a re-creation of a brook running through the edge of an English woodland, into a newly formed "rewilded" beaver pond and wet meadow. The garden used nearly ninety plant species, including many wetland species rarely seen in gardens. This seems innocent enough, a celebration of the plants of the British countryside. But the winning garden was criticized for being too laissez-faire, with not enough "muscling in" on "Nature." In this view, "rewilded" gardens are expressions of native-species puritanism that could cause a catastrophic loss of imported and cultivated varieties of plants.

It's a matter of degree, perhaps. Surely plants from other places have a place in our gardens. But they should be ornaments, not foundations. This is especially true in landscapes such as cities, where gardens offer some of the only places local species can live. Slurring advocates of the wonders of local flowering plants as misguided puritans seems unnecessary and unhelpful.

Why leaders in the horticultural community should express such forceful opposition to gardening that foregrounds or focuses on local species is a puzzle. In a time of ecological crisis, when the diversity of life is plummeting, an emphasis on local ecologies in gardens should be celebrated, not scorned. In the last ten years in the UK, half of butterfly species have declined. Two thirds of moths declined over thirty-five years, some at precipitous rates. Across Europe, one quarter of bumble-bees are on the verge of extinction. In the US, butterfly abundance fell by 22 percent from 2000 to 2020, with thirteen times as many species declining as increasing. Why oppose efforts to turn gardens into crucibles of life for these species?

I think part of the answer to the puzzle lies in the radical reorienta-tion of aesthetics and intentions involved in the new, "wilder" horticul-ture. To serve the senses and interests of many species, not just the desires of humans, fundamentally changes the purpose of a garden. This reorientation puts flowers back where they have been for millions of years, as vital nodes within webs of fruitful interspecies relationships. But gardening for a fecund and decentralized web can be very different than gardening for human aesthetics, unless our aesthetic expands to include the vitality of the local natural world. The critiques also some-times misunderstand the work involved in this new form of horticulture. Designers and gardeners are not walking away from the garden, throwing away centuries of know-how and using no "muscle." Rather, horticul-tural knowledge and work deepen and expand, reaching out to other species to ask what local flowers, birds, insects, and others really need. Often, "rewilding" involves expending considerable effort managing as-sertive plants from other continents, making sure that locals can thrive. Helping local ecologies to blossom is a very different goal from breeding another pansy variety or importing species from across the world to please the human eye.

A trip across the Atlantic to an unlikely hotbed of horticultural innovation—Brooklyn, New York—might show skeptics that this new form of horticulture is just as beautiful and involves just as much skilled human work and horticultural expertise as more traditional forms of garden design and management. Brooklyn is second only to Manhattan as the most densely populated urban area in the United States. Access to gardens and public parks is very limited, especially in poorer and Black neighborhoods. This crowdedness and inequality have spurred increased funding for public gardens and given designers opportunities to create new ways to integrate human and nonhuman needs.

Eastern Parkway is one of the busiest streets in Brooklyn for both pedestrians and cars. At its western end, at Grand Army Plaza, six major roads and six subway stations converge at the intersections of busy neighborhoods. Close to this hub, the Brooklyn Museum faces the parkway and, in front of its wide glass-covered entrance foyer, garden beds crowded with wildflowers curve across a plaza. Over six hundred thousand people walk through the beds each year as they enter the museum. Many more encounter the gardens as they walk along Eastern Parkway. Until 2023, the plaza featured only lawn arranged in strips that matched the curvature of the foyer. Katie and I lived nearby for six months in 2017 and walked past these lawn dead zones many times a week. On a few hot summer days, people sat on the grass, but mostly the beds were inert and unused, cold minimalist visual accents to the museum's frontage.

Now, after a reimagining and replanting led by designer Brook Klausing and ecological horticulturalist Rebecca McMackin, three of the five curved lawn strips are a flower lover's dream, in bloom throughout the growing season. Grasses and standing dried flowers of different hues accent the winter months. Graveled paths between these beds replace the other two strips of lawn, inviting people to get close to the flowers and sit on the low stone partitions that define the edge of each strip.

"I see my role as supporting communities," Rebecca told me by phone. "People are the main constituency, especially in urban areas. But plants, fungi, and animals, too." It shows. The museum's frontage feels animated and inviting, unlike the former corporate-plaza feel. Our senses are welcomed and we feel drawn to walk among the plants, each of which has a role in local ecology, providing food or shelter for caterpillars, nectar- and pollen-eating insects, and birds searching for insects or seeds. Most of the plants are species that lived in the area before it was urbanized, with a few from other locales in eastern North America.

"Design is about taking care of the land. And gardens are an incredibly pretty way to do that. You create a space that invites people in with beauty and atmosphere, and the dynamic, the action among all the plants."

Goal achieved. The garden looks magnificent. In the summer, yellow coreopsis flowers echo the hue of Deborah Kass's iconic yellow *OY/YO* sculpture that stands at the edge of the garden beds. In autumn, at least five different American asters shade from white to purple. Dense clusters of grasses catch the low winter light and move in the breeze. In all seasons, wandering along the garden beds reveals a continual sequence of new plant discoveries. This is the botanical version of the human culture in Brooklyn, a tightly packed community full of vitality and diversity.

Rebecca continued, her voice energized, "Beauty, and art, and delicious foods are rights. Just as much as the rights to survival and dignity. You know the old socialist rallying cry, 'Bread and Roses'? Everyone should have access to butterflies and birdsong. Gardens sometimes get put in this category of frivolous things. But public horticulture and green space are actually central to all of the issues we face right now. People's well-being, the needs of nonhuman communities, loneliness. Gardens help with all these. It's a rewarding and important mission."

I asked what it is about flowers that draws us into community with each other and with nonhumans. "They're beautiful and sexy and drama filled. In the garden we can start to notice. It's such a special, special thing and I think it's the absolute best gateway you can imagine."

My favorite time to walk through that gateway is late May, when the red columbines are in bloom. Each flower is a five-armed star, red on the outside, warm yellow inside. Reaching back from the star are long nectar-filled spurs. These ornate flowers are an exact fit for ruby-throated hummingbirds. Spurs match hummingbird beaks, anthers and stigmas dab the birds' heads, and red color catches bird eyes from afar. The flower blooms right when hummingbirds arrive in spring from their Central American winter homes.

The beauty of the columbine flower draws our senses into old stories of migration and belonging. Columbines arrived in the Americas from Asia about four million years ago. In their original Asian homes, they are pollinated by bees, and the flowers there match bee preferences: blue, with short, curved nectar spurs, and petal landing pads. But when they arrived in North America, hummingbirds offered new opportunities for partnerships. Evolution soon molded some columbines to the birds, painting the flowers red, straightening and lengthening nectar spurs, and timing the emergence of blooms to hummingbird migration.

At the museum's garden, we sit next to these blooms and experience the ancient reciprocal relationships right at the heart of the city. This is especially appropriate for a museum: living history, in action. When columbines first came to the Americas, humans had not yet evolved, our ancestors were small-brained primates making nervous forays out of the African forests. The profuse bloom of red columbines in the museum's garden calls our senses and imaginations back to a time before our own origins. The garden offers many human residents one of the few places amid the hubbub of Brooklyn to see a wild hummingbird or butterfly. It helps that Brooklyn's second-largest public park, Prospect Park, and the Brooklyn Botanic Garden sit nearby, providing trees and understory for animals that might wing to the profuse flowers in front of the museum.

"Flowers offer a portal or an invitation into understanding the natural world," Rebecca later told me as we walked through a small urban park when she visited Atlanta. "They're there to get attention. They're there to be alluring. And we are seduced, obviously, just as we should be. They're a gift of beauty, but also very much a gift of connection and recognition, weaving back threads that are broken."

Rebecca also had a leading role in the creation of another life-filled series of gardens that grow about four kilometers away from the museum, at Brooklyn Bridge Park. The park sits where the East River flows into New York's salty, tidal Upper Bay, marking the northwest boundary of Brooklyn. The two-kilometer waterfront was a commercial and shipping hub for centuries, until it fell into neglect in the mid-twentieth century. Citizen groups stopped government authorities from selling off the piers and frontage to commercial developers and, starting in the early 2000s, the area was largely reclaimed for public use.

I've been visiting the park on and off for the last fifteen years whenever I'm in New York City as a short-term resident, for work, or to visit friends and family. The co-blooming of flowering plants and people has been stunning. What was a raw concrete waterfront with disused warehouses is now a patchwork of native plantings and public amenities such as athletic facilities, picnic areas, and places to amble and sit. Large piers jut into the river, each one with a different theme—grassy knolls, dense shrubbery, covered athletic courts, playgrounds, a marina—all interconnected by walkways and gardens along the waterfront, and sound-blocking berms between the park and the expressway jammed with cars to the east.

Horticultural work at the park was first directed by Rebecca Mc-Mackin and is now led by Rashid Poulson, director of horticulture, with design by Michael Van Valkenburgh Associates. The renewal plan was implemented by a nonprofit corporation over nearly two decades. The profusion of local plants and renewal of local food webs happened not by tossing out horticultural knowledge and walking away, but through thoughtful design and experimentation. The park is mandated to be financially self-sufficient, and so renewal of the area involved building condos, restaurants, and retail space alongside the free public park. Unlike the compact museum garden, this park is large, eighty-five acres, 90 percent of which are freely open to the five million people who visit each year. Sixteen of those acres are managed as planted garden areas, mostly featuring local plants.

"The design is the launching pad." I'm on a Zoom call with Rashid.

"In the map behind me here, there's a rhythm of passive, active, passive, active throughout the park. And then continuity of the upland areas as well." He turns and points on the map at the piers and the wide walkways, then the lawn areas and planted zones. "The design naturally cues people into what the space is used for. So there's a lawn here and a volleyball court there, and every step you take to get to those places you're surrounded by this intensity of mimicked woodland edges and tons of native plantings from bed to bed, and that continues to expand." He turns back, smiling wide and throwing open his arms as he talks about the growing plantings.

Because the park is large, there's room for different feelings to coexist, often within half a minute's walk from one another. At the pier closest to the Brooklyn Bridge, small lawns for picnics and lolling are wrapped with dense planting of shrubs and trees. The Manhattan skyline across the East River moves in and out of view through these trees. Alongside, rainwater runs through five interconnected wetlands, where over 160 species of wetland-associated plants such as native grasses, sedges, asters, swamp milkweed, and small trees filter the water before it is stored underground for reuse as irrigation during dry spells. Farther south, at the water's edge, restored saltmarsh grasses grow alongside wooden pilings left over from industrial days. These grasses are nurseries for aquatic animals, habitat for ducks and shellfish, and they protect the land from flooding.

At other places in the park, walkways are lined with hedgerows of local species like wildflowers, blueberries, sumacs, locusts, and oaks. Five acres of meadows are interspersed among the woody plantings, open areas filled with nearly fifty flowering plant species such as asters, sunflowers, goldenrods, joe-pye weed, skullcap, dogbane, lobelias, iris, and sage. Flourishing plants draw in butterflies, moths, bees, nesting birds, and flocks of migratory songbirds.

Unlike much of the sterile frontage of the East River in New York

City, this park feels lush and life-filled. Even in the middle of winter, the vegetation is lively with swamp, song, and white-throated sparrows, American robins, and a few woodpeckers. All in a place that, on busy days, throngs with people. The interweave of human and nonhuman life is beautifully intense. They don't use the word *rewilding* here, but the profusion of local species and their tight weave with human life is wild in every sense. Perhaps even that misanthrope Henry David Thoreau might have sensed the life-giving power of these convergences: "In Wildness is the preservation of the World" indeed. Wild in the heart of the city. Not by accident or laissez-faire, but through design and hard work.

The fruits of that work are evident in the good human and ecological energy at the park. Also, weirdly, in spreadsheets. Rows and columns of databases don't normally make thrilling reading, but those created by Rebecca, Rashid, and their colleagues are like miniaturist portraits of the relationship between plants and their caretakers. In each, I read loving curiosity, an exploration of how to reweave the multispecies community. These are shared on the Brooklyn Bridge Park website, micro-narratives of how people and local plants can work together. For example:

Fragrant sumac: "A big problem solver. Can handle sun to shade, dogs, and salt. . . . Plants will interlace with each other to form an indistinguishable mass. . . . Larval host for two species of Hairstreaks [small butterflies]. Great fall color."

Creek sedge: "Rare in NY. Evergreen and can take light foot traffic and dog pressure. Seeds in. A real shade problem solver. Can look floppy and messy midseason."

Upland ironweed: "Smaller than gigantia and more manageable, but still massive. Will take dryer soils. Use in back of bed. . . . Great pollinator plant, dramatic, provides vertical structure to palette."

Foamflower: "A reliable, tried and true woodland ground cover. Attracts specialist bee species and other pollinators. . . . Does not tolerate foot traffic or dog pressure, but handles pathway edges and dryness surprisingly well. . . . Pay attention to cultivars."

Eastern annual saltmarsh aster: "Listed as threatened in New York State. Does well in salt marshes and meadows. Seedheads looked great in sunlight. An annual that is not reliable in the landscape. May be abundant in a certain area one year, but not the next. Moves around."

Kentucky coffeetree: "Highly ornamental bark. Beautiful, unappreciated form. Difficult to find specimens that have not been headed off at the nursery. . . . Large seedpods require cleanup on paths. Many interesting ethnobotany and history facts for tours."

Smooth blue aster: "One of the prettiest foliage. Host for pearl crescent [butterfly]. Less susceptible to Lacebug than other asters. Can get tall. . . . Somewhat dog tolerant."

. . . and on, for 250 plant species. Even mulch gets thoughtful analysis from the perspective of building good soil and giving insects a place to overwinter.

These lists and annotations are visions of a better future, one where horticulture focuses its intense, skillful attention not on the latest imported cultivar, but on the needs and glories of local plants. Many of the species on the list have seldom been used in gardens and parks, so each entry in the spreadsheet is part of a process of discovery. Human needs are accounted for, alongside the requirements of insects, birds, and threatened flowering plants. This reorientation toward local species and the needs of nonhumans is built on traditional horticultural concerns like the aesthetic qualities of flowers, foliage, and bark, the performance of each plant under foot traffic, dog pee, or different light levels, the plants' eventual size and relation to their surroundings, the best methods of propagation, and ideal placement with regard to soil, moisture, and light.

For plant people, this is gripping reading. But the granular knowledge also serves the purpose of creating wonderful experiences for people who are not horticulturalists and perhaps have never given much thought to plants, local or otherwise. Like Rebecca, Rashid got especially animated talking about how the park connects people to plants and animals, an opportunity that's hard to find in many parts of the city. He told me, "I especially remember a young girl at the Pier 6 Uplands, which is our biggest playground area. There's a small garden that's around a bog, with equisetums and all these viburnums that we keep low to the ground. It's scaled for children. The girl was talking to her family, and this was the family's first time within the park, but she'd been there before with her school." Rashid leans into his webcam and looks up, remembering with a gleam in his eye. "She's making her way into this entrance of the bog area we call the Exploratory Marsh. She turns around to her family and she goes, 'Wait, you guys, it's a forest in there.' And that clicked. That was the moment that I will never forget. I was like, oh yeah, this is this New Yorker girl's woodland. This is her meadow. This is where those gears were turning for her. This is her place to find insects, overturn stones, pick up sticks, and just simply be imaginative with them."

Later, when I met Rashid for coffee at the park, he again recounted this story, emotion catching his breath as he spoke. "My favorite thing to do here is engage with children and teach them about plants. As an urban kid, I didn't know I was missing something in my life. When I walked out of my front door, the London plane tree, the tree wells, herbaceous plants, they never really stood out to me. It was plant blindness.

The fact that school trips are taking kids to a park like Brooklyn Bridge Park, with the intention to learn about plants and horticulture, that . . . that's powerful to me."

❦ ❦ ❦ ❦ ❦ ❦

A garden is a living manifestation of our values. The formal gardens of French kings revealed their love of order, showy extravagance, and control. European nineteenth-century gardens often combined zeal to supply the kitchen, thirst for colors to defy long gray winters, and fascination with plants sent back from colonized lands. Wealthy merchants in eighteenth-century southern China showcased plants acquired through their extensive trade networks. These gardens also displayed the cultural sophistication of their owners with allusions to classical paintings, poetry, or mythology. This was especially true in the city of Guangzhou, which was a hub for flower breeding and trading, and the source of many of the Chinese flowers brought to Europe by Western horticulturalists. Modern American suburban lawns display the fruits of a diligent work ethic, helped sometimes by the homogenizing power of petrochemicals. Gardens hold mirrors to our philosophies.

Gardens not only reflect values, they create them. By constructing our sensory experience of the "natural" world around our homes, gardens shape our psyches and beliefs. A garden teaches us what nature is. Designing a garden is therefore an exercise in molding worldviews. In tending a garden, the garden makes us. It is no accident that the biblical account of our schooling in morality happened in a garden.

Agricultural areas, forests, seagrasses, and other habitats cover far more of the Earth's area than do gardens. But gardens play an outsize role in shaping how we relate to the living world. It is in private and public gardens that the majority of people actually encounter the living flowering plants whose relationship-building made our world.

The beauty and vitality that millions of people experience as they visit gardens, parks, and flower shows are therefore powerful catalysts for cultural change. To see local plants thriving in a postindustrial meadow at an elite flower show teaches us that life can spring from unexpected quarters, and we should not write off "blighted" areas. To walk through a flower- and butterfly-filled field on your way to the subway station or to play basketball on a repurposed pier edged in woodland is to have your senses enmeshed with more-than-human lives. To hear migrating white-throated sparrows sing from thickets of sumac as you take your family to picnic is to participate in the ancient pulses of the seasons and animal migration. To see a rising tide seep through spiky saltmarsh grasses is to viscerally understand that flowering plants mediate the relationship between sea and land, and will shape how the city fares as sea levels rise.

Garden design and the tending of flowers seem, at first, sideshows, perhaps even decadent distractions, in the face of the mounting crises of our world. Yet, crisis demands radical cultural change. A lived experience of thriving flowers and humans collaborating to build new life around them not only inspires that change, it embodies the answers we need: multispecies communities, working together.

Speculative Futures

Playful weeds amid ruins and the flourishing of evolution's creative powers

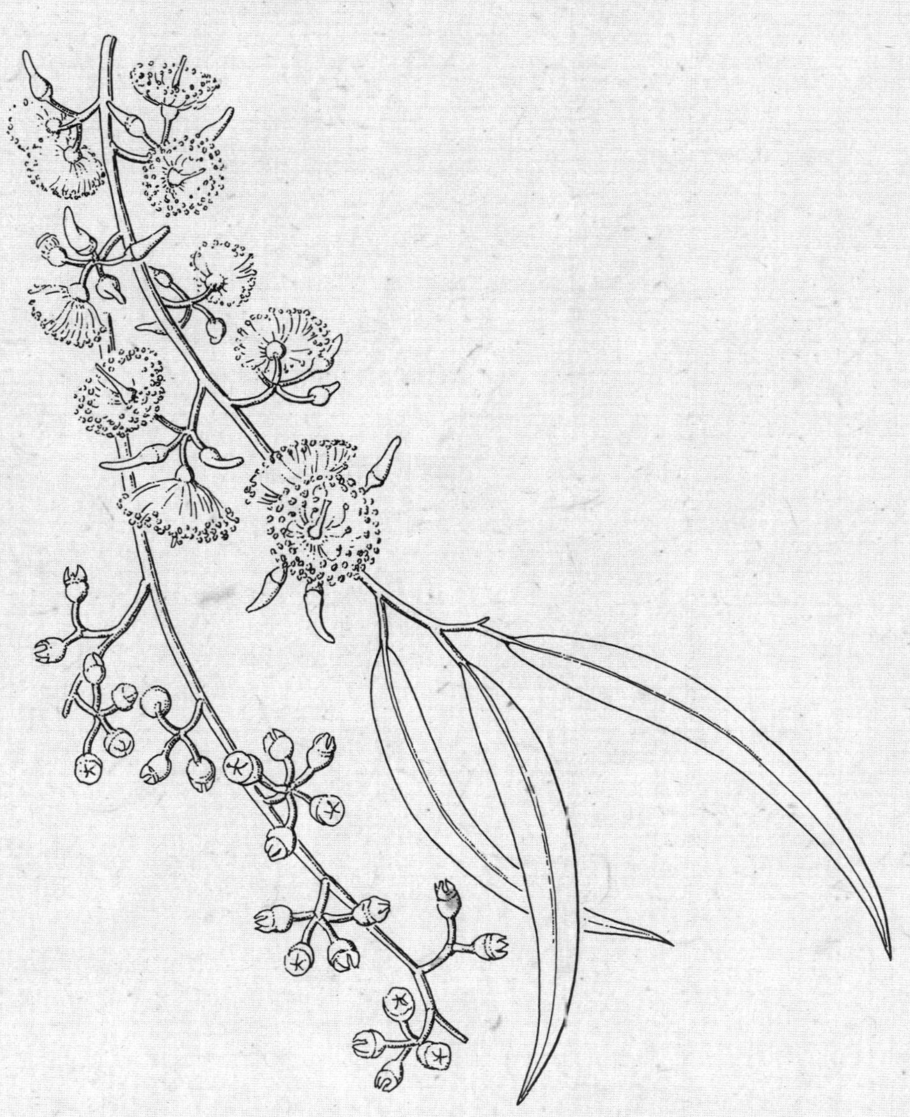

PEERING INTO THE FLORAL FUTURE IS A FOOL'S game. The chaotic intersections of technological, ecological, and cultural change raise an obscuring fog. One million years ago, could our ancestors eating wildebeest and grass seed on the African savanna have foreseen plows, fields of grain, and feedlots? Did the chrysanthemum breeders and other floral experts of eighteenth-century Guangzhou have an inkling that descendants of their blooms would populate industrialized nurseries worldwide two hundred years later? Unlikely, except in the strangest dreams.

But as the Yiddish proverb tells us, "A whole fool is half a prophet." So, let's make some opening moves in this game. I offer sketches of how flowers might continue to remake the world. I look at two very different timescales, first the short term of the next centuries, then a much longer view of tens of millions of years.

I center the flowers' agency: In what ways might their special qualities contribute to the vitality and diversity of the planet in the future? Each sketch is a speculative future, a prognosticating dream that draws on what we've learned from past floral innovations. We catch glimpses of futures both terrifying and hopeful.

* * * * *

In late winter, when I yearn for green renewal, I crouch on fissured concrete. Hairy bittercress flares from the crevices, every plant an explosion of dozens of miniature leaf stalks with paired oval leaflets ranged on each side. Warm days in late January wake them from their slumber in the cracked old driveway that runs next to our house in Atlanta. By the first week of February, the leaf rosettes are as wide as my palm. White flowers unbud, nestled at first among the leaves, then reach up on skinny supports. I bend, or sometimes lie down, and magnify the blooms with a hand lens or the zoom of a phone camera. Four unsullied white petals form a cross, veins faintly glowing at their bases. At the center of the cross, a stubby green column protrudes, circled by four filaments bearing at their tips globes of yellow pollen.

Like finely wrought metalwork enameled with colored glass, the plants are jeweled miniatures. The flowers' opening in the dreariest part of winter lifts my spirits. The gray shroud of the world will be pulled away soon, they tell me. Their appearance in a rough old driveway is especially delicious, a reminder that some flowers can flourish amid the detritus of human civilization.

There are 261 species of bittercress, all of which grow as small herbs. They are globally distributed, most growing in the woodlands, streambanks, and meadows that have been their native homes for millennia, each species adapted to a part of its home continent, such as the Blue Ridge bittercress that grows in the moist forests in the mountains of eastern North America, the lilac bittercress of the Australian Great Dividing Range, and the nine-leaved toothwort of rocky woodlands in eastern Europe. A few species, including the European hairy bittercress in our driveway, have lately vastly expanded their ranges by hopping among continents as stowaways on global trade. These species—originally from parts of Asia, Europe, or Africa—are now found all over the world, especially wherever humans have uprooted local plants and soils. These "weedy" bittercresses exult in gardens, plowed fields, and broken asphalt. Like many other weeds, they are inheriting an Earth reshaped by humans.

What is a weed? We usually define them in relation to human desires. A weed, in this view, is any plant growing where we find it ugly or inconvenient. This meaning of the word revolves around human whims or needs. Before they got turned into desirable garden plants, pansies, for example, were "noxious weeds." Periwinkle, a creeping vine with pretty blue-purple flowers, is a lovely garden plant or, when the plant chokes local vegetation in nature reserves, a weedy pest. Palmer's amaranth is today a major weed in corn and soybean fields in the southern United States, but grows naturally in southwestern grasslands and is cultivated for its nutritious seeds by Indigenous peoples.

Another meaning of the word *weed* focuses on the plant, not on its value to humans. Under this definition, the one that I and many biologists prefer for its plant-centered meanings, weedy plants are those that are good at arriving on recently disturbed soil, swiftly dropping their roots, unfurling leaves and flowers, then flinging fruits to the wind or onto animal fur in the hope of finding a new patch of exposed soil. Weedy species were hopping around the landscape long before humans evolved and started judging them. Before bulldozers and spades, these weeds sought recently burned soil, exposed areas around fallen trees and landslides, ever-shifting stream- and riverbanks, and the churned aftermaths of hoofed and burrowing mammals. Now we've remade the world in the weeds' favor.

As bittercress and other weedy plants inherit much of the green world, they homogenize the flora as they go. The species lists of plants growing in cities and agricultural fields are similar worldwide—bittercress, ragweed, horseweed, chickweed, lambsquarters, morning glory, billy-goat weed, Palmer's amaranth, annual meadow grass, and couch grass—and so, at any given latitude, one continent looks much like another, largely because of weeds.

In the month after they flower, a striking difference emerges between the bittercress on our driveway and plants of the same species growing on the street verge nearby. The two types of bittercress look, at first glance, like different species. The driveway plants hunch in mosslike mounds. They keep their skinny seed pods on short stalks. Those on the street reach gangly leaves skyward and lance pods on stems that reach two or three times higher than those in the driveway. Yet, their identically shaped flowers reveal them as the same species.

Bittercress, like many weeds, is a shape-shifter, molding itself to the particular demands of its homes. On our driveway, I periodically scythe or string-trim the plants, aiming for a semblance of order in the area that leads to the kitchen door. On the verge, neighbors mow late in the season, well after the explosive bittercress pods have shot their seeds and the plants have gone dormant for the summer. The driveway plants have learned the hard way: Keep your head low or get decapitated. Those on the verge have no such experience and are unrestrained in their reach for the light.

Such flexibility is the key to weedy success. When the world is in tumult, roll with it. All plants do this in two ways, and weeds are especially good at both. First, individual plants can reshape themselves based on their experience of the world as they grow. Second, each new generation of plants can adapt to home through genetic evolution.

When I sow seeds collected from each bittercress type in our neighborhood on the edge of an unmowed flower bed, they all grow into tall plants, luxuriating in the rich soil. This is hardly a definitive experiment, but it suggests that the driveway and verge plants differ not in their genes, but in how experience has molded their growth. After their first mowing, the unfortunate driveway plants learn, through changes in their hormones and other chemical growth regulators, to resprout as low plants, avoiding further losses. Many plants mold themselves to local conditions in this way, learning to fit in. After deer browse a bramble, the plant grows thornier shoots, the better to protect itself. Shaded plants grow bigger leaves than those in full sun, investing more of their bodies in the search for meager light. Weeds are fast growing and so can make these adjustments over weeks, not months or years as in many other plants.

Genetic change also adapts plants to the challenges of a human-dominated world. The short life cycles of weeds allow them to adapt especially swiftly. Consider, for example, the field pansies that grow around Paris. They grow as weeds in wheat and rapeseed fields, hoisting creamy yellow flowers atop long stalks. In the last thirty years, they've lost many of their pollinators, part of the widespread decline of insects across Europe. Thanks to preserved collections of field pansy seeds made by scientists in the 1990s, we can see how the flowers have adapted to the impoverished insect community. By bringing out of storage the 1990s seeds and growing them alongside those of today's pansies, the researchers found that these pansies recently evolved smaller, less nectar-rich flowers that stay open for fewer days than pansies of thirty years ago. The modern blooms still attract bumblebees, but less so than their 1990s versions. Today's flowers are more likely to self-breed, using their own pollen to fertilize eggs rather than bumblebees to bring in pollen from other flowers.

These field pansies are a different species than those bred by horticulturalists for our gardens. The wild field pansies have lately become smaller and less showy, their answer to a world with fewer insects. Their relatives in gardens have done the opposite, guided by human aesthetics to become larger and more flamboyant. In both cases, the flowers are like water, flowing through genetic evolution into the twisting ecological crevices created by human actions and desires.

Studies of weeds in urban areas around the world have found similar trends. Evolution has made many flowers less sexy than they were in the 1990s. This is a prime example of the loss of beauty and interconnectivity that happens among flowers when humans shoulder out and poison insects. At the same time, the flowers demonstrate how genetically malleable and resilient they can be.

Flowers are uniquely flexible when it comes to changing their sexual strategies. For millions of years they have adjusted their investment in floral advertisement to meet conditions. Because weeds cycle through one or two generations every year—unlike the decadal or longer generations of many other plants—their evolution is especially rapid. Tossing out showy flowers in favor of more demure self-fertilization is an old game, one helped by a host of genes that control flowering, each one ready to shift as natural selection demands. Long before humans evolved, flowering plants figured out that, although pollinators can be great for ferrying pollen and bringing in fresh genes, insects can sometimes be unreliable. This is especially true for plants that bloom where the weather is often bad, insects are rare, or plants live at lower densities. In these situations, flowers have repeatedly evolved the ability to self-fertilize. One recent scientific review called the evolution of self-fertilization "the most common evolutionary transition" among flowering plants, usually as the flowers' way to make seeds in challenging environments. These seeds, though, are inbred and usually don't grow as well as offspring made when flowers interbreed with one another. For many plants, then, self-fertilization is a backup option when insects don't show up. But for plant species that live in places where insects are unreliable, producing many seeds regardless of quality is advantageous, "selfing" becomes a permanent strategy. A few species, like many violets, even produce some flowers that fertilize themselves inside closed buds that never open to welcome pollinators. Poke around at the base of violets growing in lawns and you'll often find these self-loving flower buds at the tips of burrowing stems.

The sophistication and flexibility of female sexuality in flowers is what makes such adaptability possible. Old biology textbooks portrayed plant sex as a male hero's journey: The pollen grain lands on the flower, then grows a tube down which sperm cells gallop, as if on a knightlike quest to find an egg, trying to outrun other knights along the way. In this version of the story, the female parts of the flowers are Sleeping Beauties, passively waiting for animation by the male's kiss. In reality, female parts of the flower run the show. It is the female tissues that decide whether sperm will reach eggs.

This female agency takes many forms, depending on the species. In most flowers, she blocks any pollen from the same flower, stopping self-fertilization. In some species, the stigma, the sticky surface that gathers pollen, tastes the pollen. If she detects proteins like her own, she forbids germination. In other species, she assesses the pollen tube as it creeps through her tissues and, if it is too much like her own, shuts it down. Among a few species, the female part of the flower waits until the sperm arrives at the egg, then decides whether he's a good match. All this decision-making is supported by a host of genes that advertise and assess identity in the flower. Evolution can tweak or remake these genes, giving flowers sexual flexibility unlike any other plants.

If barriers to self-fertilization are switched off, the female parts of the flower allow pollen from their own flower to unite with an egg. This causes incestuous pairings of egg and sperm, but an inbred embryo is better than no embryo. In the past, this change commonly happened when flowering plants arrived in new habitats, like the denuded landscapes left after glaciers retreat. Now the same genetic adaptation allows flowers to make homes in places transformed by human actions. The hairy bittercress in our driveway, for example, can self-fertilize as needed. A single seed is sufficient to establish a population. Among the many bittercress species around the world, those that do well in disturbed or newly opened habitats like postglacial landscapes, fields, and cities tend to self-fertilize or, sometimes, turn off sex altogether and propagate only through bulbs and spreading stems.

Occasionally, mutations in single genes are enough to adapt a plant to a human-dominated world. Herbicides, for example, often destroy essential proteins inside cells, killing the plants. By changing their DNA, flowering plants reshape the vulnerable proteins so that the herbicide can no longer attach to them. In this way, hundreds of flowering plants can now thrive under a rain of herbicide and legions of weeds are adapted to the very poisons aimed at killing them, to the consternation of farmers. The arms race between chemical manufacturers and plants has bred what are now called "superweeds," plants so resistant to one or more herbicides that they become a stubborn presence in agricultural fields, forcing farmers to attack from multiple angles, switching or combining herbicides, using fire, or resorting to repeated tilling or hand-pulling. Nevertheless, the weeds persist.

Superweed seems an apt metaphor for the short-term future of the Earth. Some scientists and activists favor naming our present time for ourselves—the Anthropocene—but the era of superweeds seems just as appropriate, and will remain true long after we humans have gone or evolved into something else.

Loss of pollinators and rains of herbicides are hard on plants. A more dramatic challenge comes after mining. Mines not only strip away vegetation and soil, but they also often contaminate land with heavy metals. Like the biblical account of the salting of the destroyed city of Shechem or the apocryphal story of Romans salting the fields after destroying Carthage, our mining practices ruin the fruitfulness of the earth. Unlike salt, heavy metals linger and many old mine sites stay barren for centuries. Most seeds that germinate in these places are quickly killed when excess zinc, lead, or other nasty elements disrupt their cells. Even in these moonscapes, though, genetic mutations allow a few flowers to adapt.

Sea campion, a member of the carnation family, for example, has made its home on old mine sites in England, Wales, and Ireland. The plant typically grows along the coast, sending its roots into the ever-shifting crevices and sands of dunes, shingles, and rocky cliffs, then growing in low mats of silvery-green foliage. Its flowers have five cleft white petals that fuse at their base to form a swollen bladder. Sea campion seeds are the size of a piece of ground peppercorn. Some caught the wind and blew inland to the bare mine sites. Of these few, an even smaller number had genetic quirks that allowed them to sequester and neutralize excess zinc. Today, these zinc-tolerant sea campions prosper in the wastelands. Their white flowers tremble in the slightest breeze, offering nectar-rich flowers to long-tongued bees and moths, restoring the buzz of life. Flowers are also returning to mines elsewhere. Yellow monkey flowers have colonized old copper mines in California and violets sprout on lead- and copper-contaminated waste heaps in Germany. It has taken no more than a few centuries, at most, for floral beauty to return to barren land. Although scientists don't use the term in this context, it seems to me that these plants are superweeds, too. They are restoring blooms to poisoned lands, surely a superpower.

Tweaking a few genes is often not enough. Sometimes adapting to change demands an all-encompassing shakeup. The goatsbeards that we visited earlier in the book are one example. A few flowers allowed pollen from entirely different species to merge with their eggs. The resulting hybrids spread, becoming one of the most common weeds in their region, growing faster and stouter than their old-school, premerger parents. Studies of other flowering plant species—both wild and domesticated—show that merging species and doubling the genome often gives offspring an edge in tough conditions, granting them increased tolerance of salty soils, hard freezes, heat stress, drought, and other stressors. During past mass extinctions such as the calamity caused by a meteorite and volcanoes at the end of the Cretaceous, genetic revolutions like doublings surged, allowing flowering plants to find their way amid global upheaval. The ancestors of grasses, legumes, and other modern plants made it through the storm and thrived thanks to these genetic revamps. We are now in another such era, and genetic reinvention is again in vogue, especially among weeds.

As I admire the bittercress in our driveway or the goatsbeard on the margins of a wheat field, it often seems to me that these weeds are playing. Their behaviors resemble what we call play among animals: exploratory, imaginative, self-directed, rule bending, and moving in an open-ended way toward useful life skills. Weedy adaptation looks very much like spirited play. A less uplifting analogy is that these plants are the few survivors of our war on life's community, creatures whose resilience is gritty survival, not play. Against long odds, they have made new homes amid the ruins. Whichever analogy we use, the process for plants is the same: nimble exploration of new lifeways in the face of severe challenges.

Will weeds inherit the Earth? In many agricultural fields, they already have. Many crop plants are descendants of weedy plants. Remember that weeds existed long before humans, adventurous seekers of disturbed lands. Some of these species became our most important food crops. So, in an ironic twist, "weed control" in agriculture mostly involves humans trying to suppress one weedy plant in favor of another. Human value judgment of plants is fickle and often flips in momentous ways.

Wild progenitors of wheat in the Middle East are weedy annual grasses that reseed themselves after fires, churning of soil by grazing or burrowing animals, and in the unstable soils of rocky areas. In South Asia, both annual and perennial wild rice species live in open areas and wetlands, thriving in disturbed soil or wet mud. Wild maize, teosinte, is another weedy species, growing on the margins of streams and in open areas in glades and rocky slopes. Other grain species, like the ancestors of barley, oats, and rye, were weeds in every sense of the word, first sprouting unwanted in cultivated fields of wheat. These interlopers often evolved to mimic the shape of wheat to avoid being "weeded out" when farmers processed the harvest. But a few clever farmers eventually realized that these weedy grains might be quite tasty and useful on their own. To rephrase the New Testament advice about strangers and angels, "Be not forgetful to entertain weeds: for thereby some have entertained cereal crops unawares."

Much of agriculture today, then, involves humans working on behalf of weeds. We bare the soil for them and solicitously collect, preserve, and replant seeds. Weeds colonize not just bare ground, but the human psyche.

Now this old bond between humans and weedy grasses is taking new turns. We can eat the results of this cooperation today. My favorite bread flour is called Climate Blend from King Arthur Baking Company. Instead of using a single type of wheat as most flours do, the company combines multiple varieties, each of which thrives in different conditions. Developed in partnership with the Breadlab at Washington State University, not only does this flour taste great—nutty, warm, many layered—it is the result of collaboration with the genetic diversity of wheat. As the world gets hotter and more drought prone, yields from current varieties of wheat often decline, and so the search for new varieties is urgent. Instead of using single varieties of wheat engineered by humans to face climate chaos, the Climate Blend project hands over some agency to the wheat itself. Which varieties will do well this year or this decade? By planting multiple, genetically diverse wheats, farmers and scientists let the plants themselves find some answers. This is playful crop breeding.

Another place seeking to work cooperatively with flowering plants is the Land Institute in Salina, Kansas. Instead of plowing up the soil annually to plant monocultures, scientists at the institute breed perennial grains and sow them in mixtures with other plants like legumes that fertilize the soil by capturing nitrogen from the air. That's how prairies have created meters-deep soil and some of the most ecologically productive and diverse ecosystems on the planet. The institute's vision is audacious, aiming to change humanity's ten-thousand-year-old habit of growing annual grass crops into a future in which perennial grasses feed us.

I visited the Land Institute a quarter century ago, back in 2002, to see some of the first perennial grasses bred to feed humans. My most vivid memory is how ordinary these revolutionary plants seemed. To the eye, they were just regular grasses, arranged in labeled pots. Some were a little taller, greener, or stouter stemmed than others. This diversity was the secret to the plant-human collaboration. We might start the process, but the genetic diversity of the plants, sometimes enhanced by cross-breeding different species, was a form of open-ended exploratory play. Today, the descendants of some of those early varieties sit in our kitchen in a bag of commercially grown flour. They also show up in breads and beers at the supermarket, often under the trademarked name Kernza, a variety developed at the Land Institute from a Eurasian perennial grass. Did the muffins I baked this winter from these grains taste of the future? That seems a metaphor too far, but I enjoyed the richness of the muffin's flavor, a sweet taste with a slight edge of bitterness compared with the blandness of mainstream wheat flour, like the difference between an ale and a light lager.

These scientifically driven projects parallel and learn from traditional forms of agriculture. People for whom locally based agriculture on small farms is the main form of sustenance—today 84 percent of all farmers worldwide work plots of land smaller than two hectares, producing 35 percent of the world's food—have much to teach us about how to deal with climate chaos. These farmers live with extreme unpredictability every year. Today, those of us who have access to global transportation networks compensate for local crop failures with imports. We buffer our food supply through trade. As the future climate veers into new extremes, such compensation becomes less likely. What happens when everyone's grain yield falls one year? That's a question the global food supply has not had to face for decades, but which farmers in traditional agricultural contexts have dealt with for millennia. These agriculturalists understand that the way to anticipate and survive extreme unpredictability of the weather is to collaborate with the diversity and flexibility of weedy plants.

Contrast traditional seed-saving practices for maize in Mexico and Central America with industrialized practices. Instead of picking out a singular "best" variety to replant next year, as happens in most industrial agriculture, traditional practitioners select a range of seed. By saving multiple varieties, and by extensive seed trading with other farmers in the same region, genetic diversity of the maize stays high. Farmers save seeds with a range of uses—tortillas, popcorn, roasting, chicken feed, good husks for wrapping tamales, and long-term storage—and they also select varieties that do well under different growing conditions, such as their tolerance of frost or wind, their performance on different soils, and whether they grow better in the rainy or dry seasons. Alongside these direct practical connections, many farmers value diversity for its own sake, selecting some fat kernels, some wizened ones; some tinged in red, some gold or brown; a few from the big cobs, others from smaller ones.

The industrial approach to grain crops uses human ingenuity to maximize yields under stable conditions—enough rain, fuel for machines, the right herbicides and fertilizer, government subsidies, and functioning global commodity markets. If all these align, the yield will break records year after year, creating such a glut that scientists must invent new ways of off-loading the excess maize into ethanol, sweeteners, and other fruits of climatic and economic stability. Few people expect such stability in the future.

Trusting in the inherent creativity of crop plants is also part of "maslin" cereal agriculture, the practice of planting many cereal crops in the same field. *Maslin* is a word in use since at least the thirteenth century in Britain and was likely derived from the Latin for "mixture." Maslins were widespread from the Bronze Age onward in North Africa and Eurasia, and are still used by some farmers today. In North America, oats mixed with other grains were widely grown into the late nineteenth century, especially as fodder for animals. In India, sorghum was planted with maize through the twentieth century. Wheat and rice were co-sown in parts of China. Today, wheat and barley are sown together in eastern Africa. In Ethiopia, for example, multiple varieties of both wheat and barley grow together in the same fields. Barley and oats are co-planted as a spring crop in much of Poland.

A recent survey of maslin agriculture worldwide, led by botanist Alex McAlvay at the New York Botanical Garden, found that maslins increase "reliability of staple crops, especially under pest pressure, low soil fertility, inconsistent precipitation, and other challenging conditions." Farmers who plant maslins not only hedge their bets in an unpredictable world, but also slow the spread of pathogenic fungi and marauding insects, creatures that have a harder time turning into plagues when they face a range of plants. Like communities of wild grasses, those that are diverse thrive, especially under the shifting stresses of heat, drought, and insects.

How can everyday citizens in industrialized countries deepen our creative partnership with the weedy and creative nature of agricultural plants? Expanding and playing with sensory enjoyment is one way. We can develop a taste for the complexities of new grains like Kernza and old ones like those in maslins. Our visual aesthetics, too, can find new delights. In Western culture, especially, we've been trained to find weeds ugly, and to see uniform, clean-edged "amber waves of grain" and lawns as not just pleasing but morally upright. Instead, let's find beauty in crop diversity, as traditional maize harvesters still do, and see the creative potential of plants we now revile as weeds.

In the short term, the future belongs to playful weeds. In cracked concrete, poisoned soils, herbicided monocultures, and unsprayed maslin fields, weeds are thriving. When some garden designers at the Chelsea Flower Show use weeds in their plantings, they are not just highlighting plants normally disdained by professional horticulturalists, they are pointing to the Earth's future in a very literal way.

Impressive though these nimble weedy plants are, their numbers are few. Most flowering plants face a dire short-term future.

For example, although the southern magnolia species in this book's first chapter is doing well, mostly because humans love to plant the species in gardens and parks, about half of the over three hundred species of magnolia worldwide are threatened with extinction, largely because of habitat loss. The same is true for about half of the orchids for which we have enough information to judge their population status. Among other flowering plants, our knowledge is so poor that fewer than 10 percent have been assessed for the global "Red List" of threatened and endangered species. The news from the better-known groups is not good. For many, about 40 percent of species are threatened with extinction. These numbers are comparable to the much better-known vertebrate animals: 41 percent of all amphibians are threatened with extinction, 37 percent of sharks and rays, and 26 percent of mammals.

There are many horsemen driving this apocalypse, all interacting to scythe back biological diversity: habitat loss and degradation, overexploitation, climate disruption, diseases and competitors newly arrived from elsewhere, and pollution. One answer to the question "What will flowering plants contribute to the future?" is *nothing*, at least for the species that we erase. In the past, it took meteorites or global spasms of volcanism to cause mass extinctions. Now the appetites of one species of mammal are doing the same. But alongside the profound loss, weeds find new ways to thrive.

Green tendrils clamber over the ruins, garlanding the crumbling remains with flowers. There is beauty in this vision of resilience and resurgence. I certainly feel this when bittercress flowers erupt from cracked concrete. But such beauty is slippery with bad allegory. From Hubert Robert's eighteenth-century paintings of vibrant plants and hearty peasants amid fallen grand buildings, to Thomas Cole's nineteenth-century *Course of Empire* paintings, to the fake ruins built by English aristocrats on their estates, Western art has aestheticized the fall of human power back into an imagined innocent green "nature." But these titillations are attractive mostly to those not actually living in ruin. It is perhaps no coincidence that such art blossomed in Europe as colonial and industrial destruction were accelerating, reassuring viewers of nature's resilience, an anodyne for those profiting from ruination.

For the plants, though, the story does not lead to an absolving conclusion that, when humans are gone, the green world will soon restore itself to Edenic stability. There is no Romantic sublime in a future where upward of half of flowering plants go extinct in a matter of decades or centuries. Human ruins wreathed in weedy vines might be picturesque in a painting. The erasure of millions of years of evolutionary creativity is not. In the short term, we are in an age of severe diminishment, with only a minority of weedy flowering plant species thriving through rapid, playful adaptation.

* * * * * *

What happens when we vastly lengthen our view? How might flowering plants reshape the world tens of millions of years into the future? Conjecture gets tenuous at such a far remove, but we can look to present-day flowers to glimpse possibilities.

It is unlikely that humans in our current form will still be here. The fossil record tells us that most mammal species last about three million years before dying out or evolving into new species. Regardless of our fate, the effects of our present-day activities will certainly be felt for millions of years. We are carving our marks into the planet by causing mass extinction, climate chaos, and the physical and biological transformation of nearly every habitat.

If current human behavior holds, sea levels will rise by several meters and, as we discovered in the seagrass chapter, what are now coastal areas will be inundated and may become seagrass meadows. On land, the world will be hotter and fierier. The global cooling trend of the last thirty million years seems over, as are the extreme oscillations of ice ages that have dominated for at least the last one million years. We're returning the Earth to the toasty conditions that dominated much of the four hundred million years before this dip into a chilly climate. *Goodbye icy world, hello hothouse, it's been a while.*

In addition to bringing climate chaos, we are impoverishing soils over large parts of the planet. Erosion from rain and wind on agricultural and forestry land, urbanization, war, contamination, and losses due to desertification conspire to degrade the depth, fertility, and physical structure of soils. The rate of loss varies greatly, but a 2016 review by a global team of over two dozen leading soil scientists concluded that the "world's soils are under threat" and that the "global community is presently ill-prepared and ill-equipped to mount a proportionate response." A revised assessment in 2024 by a different team concluded that "most of the world's soil resources are in only fair, poor, or very poor condition, and conditions are getting worse in more cases than they are improving." One third of soils are already "moderately to highly degraded."

For a species whose religious creation stories often tell us that we're literally made from soil, we're doing a poor job. Scriptures aside, the fact that we rely on soil for most of our food does not bode well for humans in the future. What might we leave? In a few places just bare rock, the result of excessive erosion. In others, a callus of concrete. Elsewhere, our legacy will be prematurely aged soils, their nourishing minerals washed away and physical structures degraded.

How have flowers dealt with heat, fire, and challenging soils in the past?

A large dome of bare rock rises east of where I live in Atlanta. Along with several other large outcrops, it is a remnant of bubbles of magma from four hundred million years ago. Today, it rises like a whale's back from the sea of dark forest around it. The place is called Arabia Mountain, a term coined by quarry workers who named it for the hottest place they could imagine. The granite they hewed from the mountain was used for curbstone, chicken grit, road fill, and construction, including for some parts of New York's Brooklyn Bridge.

The walk from base to summit takes only fifteen minutes and passes mostly over bare, gray rock marked with distinctive silvery swirls. Fire is sporadic here because there is little to burn. Yet when the summer sun arrives, the stone heats almost fire hot, to 60°C, 140°F. Surely this is the least likely place to find flourishing and abundant plants? Yet, Arabia Mountain is a regional mecca for wildflowers. Evolution has taken a place as hospitable as a parking lot and conjured floral beauty and diversity.

Autumn is the most visually spectacular time. Tens of thousands of Porter's sunflowers all bloom at once. They stand only knee high, and each golden face is about the size of my palm. They root in shallow depressions and fissures between bare stretches of rock, growing from spring through early autumn after germinating from overwintered seeds. Porter's sunflower is not a recently arrived weed eking out a marginal existence on this mountain. Genetic data suggest that the species, and perhaps its close kin in Florida, split from other sunflowers two million to five million years ago. The species lives nowhere except in these rock outcrops in the southeastern United States. This sun-blasted pizza stone is its home.

In springtime, even more astonishing plants bloom in the persistent puddles and muddy depressions that pockmark the mountain's surface. They, too, live nowhere but in these outcrops. In the puddles, snorklewort—surely a contender for best plant name—sends a flower to the surface from a rosette of leaves rooted in the sand at the bottom of the puddle. The flower bobs between two floating leaves, tethered by a "snorkel" to the aquatic rosette. The flower is the size of a lentil. The seeds drop to the bottom of the puddle, and when the water dries in late spring, they sleep through the rest of the year until awoken by winter rains. Likewise, elf orpine, a minuscule sedum, grows only a few centimeters high and survives the hottest parts of the year as a seed. The plant germinates in muddy or sandy depressions in winter. By spring, the plants blaze starlike white flowers atop fleshy red leaves from every dip and hollow on the mountain.

Faced with bare rock punctuated with scattered depressions, evolution crafted plants that thrive in this unlikely place. The extravagance of their floral displays yields food for pollinators, and so the ancient work of flowering plants supporting animal life continues.

The rocky outcrops of the southeastern United States are tiny parts of a vast landscape. Only a handful of flowers evolved to make their homes there. What happens over millions of years on a much larger scale, where timeworn soil, heat, and fire affect not just a few hectares, but whole regions?

The moment you set foot in Australia, you can hear and see the vitality of one version of this scenario. The continent is the world's geologically oldest, relatively unstirred by the ice ages, river and coastal flooding, and mountain building that have rejuvenated soils in so many other places. The continent is also very flat, slowing the exposure of deeper layers of rock that can replenish nutrients at the surface. Many soils are therefore low in the nutrients that plants hunger for, such as phosphorus and other minerals. Fire regularly scorches many plant communities, helped by a climate that, over most of the continent, is permanently or seasonally hot and arid, albeit with swaths of humid forest in parts. The fossil record shows that fire has dominated much of the landscape for at least twenty-three million years. Fire may also have been important before that, in wetter times, all the way back to the Late Cretaceous, more than sixty-six million years ago. Many flowering plants encourage these burns by growing abundant fuel in the form of flammable leafy and woody material. Like the grasses at Panola, these plants kindle and stoke their own blazes, the better to regenerate.

The special qualities of the Australian flora, and the ways that these flowers catalyze animal life, struck me within an hour of staggering, jet-lagged, out of Brisbane Airport. Katie and I visited as part of my book tour, a journey that took us to parts of Queensland, down the east coast, inland to Canberra, then to Melbourne. Alongside the delights of meeting Australian book lovers, we used the visit to meet and study plant and animal life. Our visit encompassed just a small fraction of the continent's biological diversity, but enough to transform how I think about Earth's possible distant futures. It started at a roadside inn near the airport.

Waiting for our sandwiches on the front porch of the Shelley Inn, overlooking a bay north of Brisbane, our ears were jolted by strange bird sounds coming from the top of a giant tree across the road. Fluting whistles combined with tones like sweeps of an electronic synthesizer. Ratchet-like clicks trilled through these richer notes. At least a dozen birds stitched together their flock with these lively sounds. The tree, a Moreton Bay fig, was a giant, its trunk wider than the large picnic benches around it. The branches spread across at least fifteen meters and were thick with leathery foliage, blocking our view of the birds. We pulled out binoculars and eventually spied Australasian figbirds, stout-beaked, long-tailed birds about the size of American blue jays. They plucked at a scant few fruits in the tree canopy. Some of the birds, females, were streaked in subtle browns. The males had olive-lemony backs, black-gray heads, and bright red "spectacles" of bare skin around their eyes. The birds' clamor was not subtle.

This first taste of the exuberance of Australian plant and bird life—garrulous birds feasting on the fruits of spectacular trees—was soon overwhelmed by the arrival of a flock of about twenty rainbow lorikeets. They live up to their name with blue heads, red-and-orange chests, green backs, yellow underwings and tails, and crimson beaks. The lorikeets arrived like windblown confetti, a tumbling fuss of wings as they settled. Each bird excitedly belted out what sounded like a cross between a child's yelp of delight and a squeaking door hinge. In unison, the chorus was loud enough to swallow the figbirds' sounds. This storm of bird color and sound was short-lived. After hopping and swinging among the upper branches of the fig tree, the flock lifted as one and winged south, along the bay front, calling as they flew.

These birds were a thrilling welcome for Katie and me. Our ears are used to the gentle warblings and relatively quiet calls of North American and European birds. We knew immediately that we'd arrived in a place where evolution had taken a different path. Birds grabbed our attention but, ultimately, flowering plants were the primary cause of the differences that we admired. We soon found out how. After lunch on the porch of the inn, we drove inland from the bay, aiming to head north to where we had booked a room. We got only one hundred or so meters along the road before I abruptly pulled to the verge. We'd caught up with the lorikeets. Not just the twenty that had briefly lighted in the fig, but more, perhaps fifty in all, swarming a tall eucalyptus tree. The sound was vigorous enough to fill the car, calling us out of the bubble of metal and glass.

Such excitement. The birds were tumbling over one another as they grappled the thin branches of the tall, smooth-barked eucalypt, a species known as forest red gum. Unlike the dense foliage of the fig, this tree had a thin canopy of drooping slender leaves, allowing a clear look at the lorikeets as they gathered at clusters of flowers along the branches. The birds shimmied and stepped on twigs, then hung upside down to stuff their beaks into frilly white blooms. Through binoculars, each flower looked like a many-tentacled sea anemone, a ring of skinny fingers ranged around a central bowl. These filaments are the male parts, carrying pollen at their ends. As with all eucalypts, we could see no petals—these are fused into a nipple-like cap that covers immature blooms—and the filaments' bright creamy color are the flower's only visual display. We could see that many of the birds' heads were dusted with pollen, like confectioners' sugar on blue cupcakes. The lorikeets were feasting on nectar, using brush-tipped tongues to slurp the sweet reward from the interior of each flower's cup. As they fed, they also gobbled some pollen, an important protein complement to their sugary diet.

The fig tree runs its life in ways similar to flowering plants worldwide. Insects pollinate the flowers (wasps in the case of the fig) and birds carry away the fruits. The eucalypt instead offers generous nectar to birds. Because the tree was so large and hung with hundreds of blooms, the resulting avian excitement was loud enough to be heard from a closed-windowed moving car.

We soon learned that many other local plants also provide abundant nectar to birds, especially honeyeaters and parrots, birds that often specialize on nectar. Imagine if plants in the woodlands of North America or Europe were seasonally festooned with nectar-rich bird-feeding flowers, and that dozens of bird species had evolved to enjoy and serve these fonts of sugar water. Instead, these other continents have forests of wind-pollinated oaks and maples, with a few insect-pollinated lindens and hollies. Birds there focus on plucking insects, seeds, and berries, usually without the ruckus evoked by a sugar-gushing tree.

We visited in August, during winter and not prime flower season, yet floral diversity and its accompanying birdlife were everywhere. Honeyeater birds—of which Australia has seventy-five species—perched on the stout, columnar blooms of *Banksia* shrubs and trees, probing for nectar between the flowers' dozens of neatly ranged yellow- and red-colored fingers. Larger parrots jostled with lorikeets in eucalypts and *Banksia*. Bottlebrush trees offered flowers like explosions of crimson thread, attracting silvereyes, small birds with striking eye rings, and a few thornbills, tiny olive-yellow birds with pincerlike beaks. Most wattle and *Grevillea* plants were not in bloom, but we spied a few of their colorful spiderlike flowers in irrigated gardens, with honeyeaters and lorikeets in noisy attendance.

"Oh my god, this is unbelievable" was a phrase Katie and I repeated many times. Our astonishment was partly the enthusiasm of visitors encountering new experiences, but also reflected the sense that, after nearly a century of combined experience of birds and plants in the Americas and Europe, we'd stepped into a story with a different narrative. Were we falling into the colonial mindset of exoticizing the lives we encountered? We are not immune to such subconscious biases. Yet scientific studies of flowering plants in Australia show that there is a special braid of biology and geology here, one that contrasts with much of the rest of the world. And perhaps foreshadows the distant future for other places.

Part of this story lies in the soil. For plants rooted in impoverished soil, growth of new leaves is constrained by shortages of minerals. Like a house construction crew with no nails, the plant may have plenty of sunlight, air, and water, but without essential mineral supplies, little building can happen. In Australia and in other places with poor soils, plants therefore often grow slowly and hold on to their leaves for longer than most other plants, sometimes for years. Strangely, even amid mineral shortages, sugar can be easy to make. Sugars have no minerals in them and can be crafted from water, air, and sunshine. Plants rooted in poor soil can therefore often pump out sweetness.

What to do with sugary excess? Some plants, like acacias, ooze it out of pores on their trunks to reward bodyguard ants that fend off leaf-eating insects. A few eucalypts offer "manna" from their bark, crumbly clumps of sugary solids beloved by many birds and an important food in some Aboriginal cultures. Other plants seep sugars into the soil, feeding fungi and bacteria, a good strategy when these creatures can reciprocate with phosphorus and nitrogen. Australia has the largest proportion of plants that cooperate with root-associated bacteria to gain the nitrogen needed for growth, a prime example of how flowering plants solve challenges through teamwork. Some plants also use looping chemical reactions to burn off sugar without affecting the rest of the cell, like pressure-release valves venting steam from a boiler. For many Australian plants, the answer to sugary abundance is to offer copious nectar to birds. In this way, flowering plants forge deep partnerships with feathered pollinators, often using the birds' favorite colors—red and orange—to attract attention. A few flowers specialize on working with mammals such as possums and the large bats called flying foxes, attracting them with nectar secreted at night.

Birds responded to this nectar-rich world with a burst of evolutionary expansion. The ancestors of all songbirds—the small perching birds that comprise half of all bird species worldwide—evolved in Australia. Surrounded by sweetness, they evolved a new taste receptor, one tuned to sugar. This strengthened the relationship between many Australian birds and flowering plants. When some descendants of these ancestral songbirds left Australia to spread around the world, they carried with them the sugar taste receptor. The fondness for sugary flowers among African and Asian sunbirds, Hawaiian honeycreepers, and American orioles and tanagers, along with the enthusiasm for sugary fruits among many other songbirds, has its origins in the soils and flowers of Australia.

As we walked through towns or nature reserves, Katie and I found birds by listening for the fuss haloing individual trees, the ones in full bloom. Because the timing of each plant's flowering differs slightly from others in the same area, birds flock to the trees with the most nectar. The ensuing squabbling raises clouds of sound, as do the cooperative cries of flock mates calling to one another to advertise discoveries. Of all the continents, Australia has the most vigorous and diverse bird soundscape, a direct result of floral evolution. For human newcomers, the experience can be overwhelming—so much energy can be energizing or enervating—and for Australians the sounds give a strong sonic signature of home, an internal compass that stays with many for life. Play the sounds of lorikeets to an Australian émigré in London or New York and you may evoke smiles as memories resurface.

Fire also promotes floral diversity and specialization. The Australian continent is, in many places, hot and seasonally arid, with frequent lightning strikes and a history of human-set fires dating back tens of thousands of years. Nutrient-poor leaves and wood tend not to get eaten by insects or fungi, and so they accumulate, building up fuel. More, many Australian plants defend themselves with oils and sugar-derived chemicals that are cheap to make when the plants have excess sugar and few minerals. These defensive chemicals are often highly flammable, such as the oils that suffuse eucalyptus leaves. When the fires come, the burn further impoverishes the soil, sending nitrogen into the air and causing minerals to blow or wash away in ash. And so plants must adapt to the specific mineral deficiencies of their homes, growing specialized root systems or inventing new symbiotic relationships with local fungi or animals. Such local specialization encourages species to fragment, increasing species diversity over time.

The paradox that plants bloom in profusion where conditions seem dire becomes especially evident in extreme cases. In southwest Australia, and also in the Cape region at the very southern tip of South Africa, soils are exceptionally ancient and the land is regularly torched. These places are home to some of the world's most diverse and unique communities of flowering plants.

At least eight thousand species of plants live in southwest Australia, and most are found nowhere else. Ancient soils, some millions of years old, dominate the region, mixed with younger areas mostly at the peripheries. Fire is common. Shrubby vegetation dominates, although in some parts, where rainfall is higher, tall eucalypts grow in dense forests. Although Australia is home to some of the showiest flowers on the planet, it is in the southwest that the diversity of flowers is most extraordinary. Eucalypts, *Banksia*, bottlebrush, *Hakea*, waxflowers, smokebushes, basket flowers, *Grevillea*, and other trees and shrubs all proffer flowers with arresting hues and strange shapes, as if a skilled glassmaker had been commissioned to explore the creative boundaries of floral form. Between the shrubs and trees, orchids, kangaroo paws, lilies, and hundreds of other herbaceous flowers add to the profusion. The association of birds and flowers is especially pronounced in the region, and the world's only terrestrial mammal that eats nothing but nectar and pollen, the honey possum, lives exclusively in the region, feeding on local *Banksia* and other flowers.

Across the Indian Ocean from southwestern Australia, the African Cape Floristic Region is an area of ninety thousand square kilometers on the southern tip of the African continent. It is home to nine thousand species of plants, nearly 70 percent of which are found nowhere else on Earth. Even though the Cape is in the temperate zone—about as far from the equator as Atlanta, Sydney, or Santiago—the richness of species is equivalent to that of a tropical rainforest on the equator. The region has experienced little tectonic activity in 140 million years, and most of the soils in the region are infertile, especially those made from eroded sandstones. Rainfall is highly seasonal, especially in the western part of the region, and fires burn regularly. Researchers in the 1980s wrote that after a fire, it "is hard for an observer to imagine that the blackened waste . . . is merely a stage in the cycle of regeneration and development of such rich and attractive communities." Yet the vegetation there, especially the shrubby heathland community, requires regular burning to thrive.

Flowers are so abundant in the Cape that flower viewing is a draw for tourists, and flower harvesting is a major part of the local economy. The blooms include the anemone- and pincushion-like flowers of *Protea* plants and their kin. Over three hundred species of these shrubby species live in the region, 96 percent of which are found nowhere else. Hundreds of other shrubs and small tree species also crowd the Cape, including rooibos, known worldwide for the tea made from its leaves. Herbaceous wildflowers like asters, legumes, orchids, and irises also bloom in profusion. As in Australia, many of these—perhaps three hundred species in all—depend on birds for pollination, and the flowers offer copious food to the sunbirds and sugarbirds, curved-beak nectar specialists that are especially abundant in the region. These plant communities may not be as lush and productive as a deep-soiled prairie or tropical rainforest, but the number of flowering plant species and their tight relationships with animals, especially nectar-drinking birds, are staggering.

When scientists map global "hotspots" of biodiversity, both southwestern Australia and the South African Cape light up on the charts for the richness and uniqueness of species. Floral evolution has turned these seemingly inhospitable lands into places that rank among the most diverse on the planet. Western science is getting a glimpse of what Indigenous peoples have known for much longer, namely that places with ancient soils and hot and fiery climates are worthy of veneration. For example, the Noongar Aboriginal people have lived in southwestern Australia for at least forty-eight thousand years—since the beginning of creation in their cosmology—and revere the places within the landscape that have the oldest soils. But my wording here separates landscape from humanity. In Noongar culture, humans, nonhuman life, land, and water are all manifestations of the same undivided story where rivers are mother's milk and mountains are heads, always watching.

The flourishing of plants and animals in worn-out soils amid frequent fire upends the stories we read into nature, especially for people like me from northern hemisphere cultures. In places like North America and Europe, where most soils are geologically young, many plants thrive only in fertile soils, protected from fire and excess heat. We expect living abundance where soils are deep, dark, and well watered. This is the pastoral ideal and the Edenic garden. In timeworn or washed-out soils, we imagine barren "wilderness" or postapocalyptic ruin. This may be accurate in the immediate aftermath of dust bowls and other degrading spasms of human improvidence. But given millions of years, flowers will not be so easily denied. It is their genius to make the world productive, diverse, and life filled, usually through odds-defying acts of physiological nimbleness and cooperative work with others.

What does life do when things get hard? Strew the world with flowers. Challenging conditions ignite evolution's creative powers.

When our spirits are beaten down, when human destructiveness seems to know no bounds and despair floods in, it is good to remember this. We humans are causing a mass extinction, a collective act of stupendous harm and disrespect for the living Earth. But we will not have the last word. Flowers made our world, and they will remake a new one when the human frenzy of biotic annihilation is spent.

Does this excuse us from our moral responsibility to care for the world now? Emphatically, no. For at least two reasons.

First, the far-off botanical glories I describe here are speculations. Speculation can give us a wide lens, sometimes illuminating, but often fogged or distorting. There is no guarantee that millions of years of future evolution will turn post-human wastelands into places of floral beauty and rich plant-animal relationships. Our destruction of so much plant life today makes such renewal less likely because it erases the raw material needed for future evolutionary innovation. This is true even in southwestern Australia and the Cape, both of which are severely threatened by clearing for agriculture, the spread of diseases and plants from other continents, and rapid climate change. By eliminating diversity, we're stymieing flowering plants' creativity and resilience.

Second, the fact that beauty and diversity sometimes reappear in beaten-down places does not justify the beating. People in war zones understand this. Poppies blowing between crosses in Flanders fields and fireweed blazing from the rubble of blitzed London are not arguments for the goodness of war. After all, if we value flourishing habitats, we should protect the ones we have, not destroy them while anesthetizing our morality with imaginings of far-off possible renewal.

Last, and most important, our human path does not have to be one of destruction. This is the flowers' most important lesson. Thriving worlds grow from cooperation, mediated by beauty, with some illusion thrown in because life's story is always edgy. We humans can learn from flowers and do the same. We already are, in some places. Seagrass restoration, grassland renewal, horticulture that honors people and plants alike, and agriculture that embraces the innovative nature of both plants and traditional cultures are all founded on floral-human creativity. We can expand and elaborate these improbable mutualisms in gloriously fruitful ways. Flowers tell us: Together, we can remake the world.

Afterword

On Floral Beauty and Joy

WHEN WE TAKE PLEASURE IN A FLOWER, WE reclaim our inheritance as animals in a world made by flowers. Sensual indulgence and enjoyment of floral beauty are not frivolities. Flowers give us embodied experiences of the creativity and vitality that drive the grand narrative of life on Earth. Floral delight is deep, earthy beauty.

Such floral power is embedded in scriptures. When the Buddha held up a flower as a silent sermon or King Solomon sang, "I am the rose of Sharon, and the lily of the valleys," these visionaries were not just offering metaphors. The flower *is* the teaching. The flower *is* the lover. These stories tell us that flowers are not passive symbols, but active, creative agents.

The same message is embedded in rituals and other cultural habits. A bouquet for love. Flowers on a grave. A shower of petals for newlyweds. Homes and public spaces circled by flower beds. Floral garlands and altarpieces in places of worship. Flowers appear at nearly every significant human life transition and gathering place. We may at times dismiss flowers as superficially pretty or box them into narrow symbolic roles, but their presence at the center of acts of love, grief, community, worship, and cultivation reveal that, deep down, we understand the central, life-giving importance of flowers in our world.

One effect of flowers on humans has special significance for our present time: their ability to evoke beauty and joy. Many of us feel alienated from the living Earth. This is exacerbated by the ravening appetites of electronic algorithms and other seekers of control that yearly get more skillful at drawing not just our attention but also our wills and imaginations away from home and kin, both human and nonhuman. Such disconnection is desolation. Beauty and joy are antidotes.

We have our ancestors to thank. They gave us bodies and minds that respond to flowers with delight, a form of inherited wisdom scribed by evolution into our nerve endings and brain chemistry. In the presence of flowers, our senses glow. Beauty awakens inside us. Sometimes this aesthetic response shimmers on the surface of consciousness, then dissipates, a momentary flare of pleasure. At other times, sensual thrill resonates with memory, emotion, and reason, igniting joy.

The joy of flowers is a gift, a form of biological grace that arrives whether or not we deserve it. Floral joy makes us more open to connection beyond the self. We're freed momentarily from the husk-like ego. Creative possibilities expand. Life is affirmed. Next time you're in the presence of a gorgeous bouquet or a meadow of wildflowers, let your senses linger. Flowers will reveal their power to remake us from within.

The joy we experience from flowers is a modern manifestation of their ancient power to speak to animals in the language of beauty and to build new possibilities. Given that humans now hold sway over the planet's fate, the texture of our psyches—including the generative, connective effects of flowers on our minds and emotions—matters to all other life.

Illusion often trails and shadows beauty and joy. A little illusion re-enchants the world, a perfume that elevates as it beguiles. Too much illusion, though, can harm. Pretty flowers grown by poisoning people and land, for example, are harmful deceptions, a form of brokenness that we must remedy. The power of these illusions is a reminder of how important experiences of flowers are for us as a species. Few things renew and motivate us more than deeply felt experiences of beauty that draw us out of ourselves. This is a good reason to give flowers our close and caring attention, seeking real beauty amid vapors of illusion.

And so say *yes* when the bartender offers a flower. Cultivate flowers and their appreciation in your life. Offer the gift of floral celebration to others. Renewing our connection to life's creativity and vitality may be flowers' most world-changing gift of all.

Acknowledgments

THIS BOOK WOULD NOT HAVE GERMINATED, LET alone bloomed, without a community of support as rich and mutualistic as the pollinators, fruit dispersers, plant companions, and underground microbes and fungi from which all flowers gain their lives. I am especially grateful for the brilliant guidance of my editor at Viking, Terezia Cicel. In the UK, Katherine Cowdrey, Susanna Wadeson, and Sally Wray at Transworld provided wonderful insight and support. Terezia and Katherine's editorial suggestions greatly improved the book. I also thank Jennifer Tait, production editor, Maya Petrillo-Fernandez, editorial assistant, and Susan VanHecke, copy editor, for their excellent work. Alice Martell, my agent, helped me to shape the book and is a peerless advocate and friend in our work together on this book and for all my previous books. I thank Lucy T. Smith for creating the marvelous art at the start of each chapter. I encourage readers to explore her portfolio, courses, and book, *Botanical Sketchbooks: An Artist's Guide to Plant Studies*. I also thank the design, marketing, and publicity teams at Viking, Transworld, and publishers of translated editions. A book, like a flower, may bear a single name but is the product of an extraordinary community. Thank you.

Katherine Lehman, my partner, reminds me every day of the extraordinary beauty in our world, if we just take the time to notice and appreciate it, especially among the flowers. Her comments on early drafts of the book elevated the text and helped me to see stories more clearly. My sister, Julia Grindley, and brother-in-law, John Grindley, are welcoming hosts when we're in Scotland; Julia made our expeditions to find eelgrass fun and enlightening, and she contributed wise advice about earlier book drafts. My late father, George Haskell, gave me a lifetime of encouragement, and he contributed some excellent ideas as I was developing the initial plans for this book. When asked why her plants thrive, my mother, Jean Haskell, laughs and disavows responsibility. In this, she is like many "plant people," among whom expertise often strongly correlates with humility and a wry sense of humor. Those of us who know her best understand that it is loving dedication that causes life to sprout and flourish around her. Julia and I are lucky to be among those sprouts. She also gave valuable comments on the manuscript and shared her deep floral experience and curiosity as I worked on each chapter.

I offer many thanks to the thousands of scientists whose work I drew on as I wrote the book. I hope that readers will explore the bibliography and use it as an entry into the marvels of the scientific literature on flowers. I'm grateful to the community of plant enthusiasts, scientists, and conservationists who generously shared their time and insights with me, including: Walter Fertig, Marion Ownbey Herbarium, Washington State University, for his welcome and expertise about *Tragopogon* in eastern Washington. Pam and Doug Soltis, University of Florida, for sharing their wealth of knowledge about *Tragopogon*. Ken Childs, for generously sharing information and photographs of moths with their eyes gummed with cranefly orchid pollen. Gabe Andrle, Melanie Furr, Sebastian Hagan, and Logan Jones at Birds Georgia, for including me in some of their work on grassland restoration and for showing how the ancient bond between grass and fire plays out in a modern context. Deborah McGrath and Jon Evans, University of the South, for many delightful and helpful conversations about plant diversity and physiology. For their generous sharing of knowledge and time, both through email and in the field, the leaders of Restoration Forth in Scotland, especially Lyle Boyle, the Ecology Centre; Addie Dinsmore and Naomi Arnold, World Wildlife Fund; and Esther Thomsen, Project Seagrass. Christine Saillard and Christophe Mège at the Musées de Grasse, Grasse, France, for discussing with me the brilliant work they do nurturing wonder and curiosity about aromatic plants and local ecologies. Isabelle Charmantier, the Linnean Society of London, for showing me some of Linnaeus's manuscripts and plants, and for generously sharing her insights and expertise about the intricate webs of connection among ideas, books, and preserved biological specimens. Colleagues at Black Inc., along with the Byron Writers Festival, the Bendigo Writers Festival, the National Library of Australia, and Integrity 20 at Griffith University, for generously hosting me in Australia. Dror Burstein and Dani Nadel for kindly answering my questions about *Salvia* and other plants in Israel. Rashid Poulson, director of horticulture, Brooklyn Bridge Park, and Rebecca McMackin, ecological horticulturalist, for enlightening and inspiring conversations, and for the extraordinary gardens and parks that they create. Fraser Woodburn, University of Bristol, for advice on floral electrical fields. Bill Keener for his many insights about grasses, agriculture, and human history. Peter Brannen, Greg Priest, and Martha Stevenson for clarifying my thinking about biological "revolutions." The Alderson-Tillinghast Fund at the University of the South supported some of my travel as I researched the book, and a sabbatical leave from the university helped me to complete part of the manuscript. Conversations with Paul Kvinta gave me much encouragement and opened new ways of thinking. Annie Novak, Marianne Tyndall, Joseph Bordley, and Sanford McGee generously shared plants, green insight, and friendship. Fellow gardeners at the Emory Grove Community Garden in Atlanta, Georgia, and the Foothills Community Garden in Boulder, Colorado, shared the joys and frustrations of interweaving human care and attention into the world of flowering plants and embodied the idea that human and plant communities thrive in mutualistic relationship. Our family's plants and cats shared their wordless creativity and inspiration, and their lives and spirits partly shaped these inked symbols.

Thank you, reader, for sharing some time with me among the flowers. May flowers lead us all, as they have for millions of years, to more delight, curiosity, connection, and fruitful, creative lives.

Supplement

Invitations to Play with Flowers

PLAY IS A FORM OF JOYFUL CURIOSITY. IT UNITES senses, intellect, and emotion. It can draw us into community, both with other people and with the more-than-human world. The activities that follow invite noses to explore, fingertips to discover, and eyes to seek out detail. I hope they also open imaginative space. Most of the invitations require minimal or no equipment, and can be done at home.

Some of the activities are brief, others are more involved. Skip around. Play. I hope that you'll invite friends and family to join you. By drawing others into the experience, you help flowers do what they do best: create connections that stir creativity.

Plant and connect

Planting local flowers in a window box or garden bed unites us with flowering plants like no other activity. By nurturing seeds, bulbs, or baby plants, we can join the ancient plant-animal friendship, tussle, and source of delight, food, and medicine. But don't go it alone. Flowers also knit us deeper into the human community. Find your local native plant society, botanists' club, community garden, or other hub of flower-growing, veggie-tending, and soil-under-the-fingernails people. Find ways of receiving and passing along plant knowledge. Invite some kids into the process and learn from them, too. Repairing the frayed connections among people, especially among human generations, is part of what growing flowers does for us, restoring culture within and beyond our species.

Press a flower between the pages of this book

Drying flowers between sheets of paper is a time-honored way to remember a moment and to interleaf your story with that of a book. Botanists use the same method of pressing flowers between paper to preserve their important findings in herbaria, usually with large paper sheets and heavy-duty presses in drying ovens. At home, all you need is a flower, a book, and some time.

Slip the flower between the pages, perhaps to mark a memorable passage or to remember the day that you found the flower. Close the book tight and wait a few weeks. If you want to get more professional, pile a few heavy tomes on top of your book. If you don't want to stain the book's pages, use scrap paper to enclose the flower before pressing it inside the book. A book squeezed between others on a shelf also works. Thinner flowers preserve best. Those with chunky, wet cores like dandelions or big orchids sometimes mold and often end up looking more like roadkill than gorgeous floral mementos. White colors lose some of their pearliness, but red, purple, and orange flowers emerge from their hug between the pages with vibrant colors.

Once they are dried, you can glue the flowers to a paper backing, then put them on the mantel. Sunlight will quickly fade many flower colors, so keeping them shaded helps to preserve the pigments. Or sneak your pressed flowers between the pages of a friend's book to surprise them. Who can fail to be charmed by the discovery of florid prose?

How to read flowers

Rose

Let's start by plucking apart a rose. The bloom will show us the themes that unite all flowers. I always feel squeamish about destroying such a beautiful creature. But the exercise is one of deepening connection, not dishonoring the plant. Once your hands and eyes have grasped the floral anatomy, the marvels of diversity among different species will be more apparent.

A ring of five leaflike projections splay where the rose's stem meets the flower. These are the sepals and they often have little spikes along their edges. When the flower was in its bud, the sepals covered the embryonic tissues within. Now they form a vase, supporting the rest of the flower. Like many of the other parts of the flower, the sepals are modified leaves. When a plant senses that the temperature and light conditions are right, and that the plant is big enough to support a flower, hormones divert some embryonic cells from making leaves to building flowers.

Inside the ring of sepals grow the colored petals, each one also a modified leaf. Wild roses have five petals, with the sexual parts exposed at their center. But we humans want more. Natural genetic mutations caused a multiplication of the five-petaled whorl, sometimes into dozens or a hundred or more petals. These are the varieties that our ancestors in Asia and Europe selectively propagated, giving us modern domesticated roses.

If you tear a petal and look closely at the ripped edge, you'll often see that the pigment is on the upper and lower surfaces, leaving a whitish interior. The petal surfaces also release aromatic oils. The fresh face of the petal is, of course, ephemeral. After a few days, and especially if the flower is pollinated by a visiting insect, cells in the petal start to die. They don't just fade away; they actively digest themselves and send the resulting nutritious slush back to the rest of the plant for recycling. This dying is coordinated and accelerated by a gas, ethylene, that petal cells waft in minute quantities over the flower, a botanical death perfume. We cannot smell it, but ethylene gas is the cue for the petals to exit the stage and allow the flower to concentrate on the next stage, maturing into a fruit.

Once you've plucked your rose, you'll have a pile of petals to shower over your beloved or sprinkle in your bath. You'll also see the two inner rings of the flower. Just inside the petals, a mop of filaments bears the orange or yellow anthers, the knobs in which pollen grows. When an anther is mature, it splits open and sheds the pollen.

Inside the mop, at the very center of the bloom, more struts, called styles, poke up, each one with a creamy cap. Peer through a magnifying lens or strong reading glasses and you'll see that each cap is roughened with little bumps and looks slightly moist. These caps are the stigmas, the sites where the female part of the flower gathers pollen.

Like the sepals and petals, these male and female sexual parts are also made from modified leaves. Early in the flower bud's development, hormones instruct the embryonic cells to make pollen or stigmas, not flat green surfaces. Plants are masters of repurposing existing forms. So, too, are horticulturalists. In some cultivated varieties of roses, most of the anthers have been turned into petals. In these roses, the mop of anthers is transformed into a crowd of petals, with just a few sexual parts at the very center.

Take a sharp knife and carefully cut through the center of the rose. Don't worry about getting it wrong—any cut works, lengthwise or crosswise or both. The swollen area at the base of the female sexual parts is a cup containing the ovaries, which hold unfertilized eggs. If your rose was immature—harvested from the greenhouse or field when it was still closed—this cup will be an unimpressive bump at the base of the flower. If your rose is more mature or, even better, is one that you brought inside after bees visited, you'll see a larger swelling and, sometimes, the developing seeds. Left alone on the plant, this cup would swell into a rose "hip," the plant's fruit enclosed within the cup, and eventually fill with mature seeds.

Four whorls of modified leaves: sepal, petal, filament-and-anther, stigma-and-style. This is the design that evolution has modified into hundreds of thousands of different floral forms. Once you know the starting formula, you can admire the inventive ways that plants have used it to fit their particular purposes.

Lilies

Lily sepals and petals look alike, both sporting the same colors. But the sepals still do the job of enclosing the bud. Look at a barely opened lily at the florist's shop and you'll see how the three outer sepals cover the three inner petals. Lily anthers are large and burst open to shake pollen all over the inside of the flower. Pull off a petal and, at its base, you'll sometimes see drops of nectar. The stigma protrudes from the flower's center and its sticky covering is often obvious. Cutting into the ovary, you can see how the lilies, unlike messy roses, arrange their eggs and seeds into three tidy compartments. This division into three (note that the sepals and petals were also in triads), or multiples of three (six, nine, etc.), is a characteristic of the larger clan to which the lilies belong, the "monocots" that also include orchids, grasses, seagrasses, and bananas.

Orchids

Like lilies, orchids also color their sepals and petals in similar hues. They convert the lower petal into an ornamented landing pad and chute for incoming insects. The male and female parts are fused into a single column that juts above this lip. A later activity shows you how to pollinate this delightfully specialized sexual arrangement.

Sunflowers

Sunflowers, daisies, and asters add another twist. What looks like a single flower is in fact a gathering of many. Lay any of these flowers onto a cutting board or hold them in your palm. Cut or pull the central disk. It will break apart into dozens or hundreds of little pieces. Each piece is a flower, a miniaturized version of a rose or lily, pressed up against others in the disk, usually with such careful shoulder-to-shoulder alignment that they form graceful spirals. Every tiny bloom, usually called a floret, has male and female parts sprouting from a tube made of fused petals. A magnifying lens helps to see each floret in its wondrous completeness as a flower. In sunflower florets, the anthers mature first, shaking their pollen from the disk's face, then the stigmas poke up, each like a forked tongue, tasting the air for pollen. The sepal is buried at the base of the floret, a dry tuft. In some sunflower relatives, such as dandelion, this sepal tuft grows into the parachute that carries the seed aloft.

The inner florets are ringed with what look like big, brightly colored petals. Each one of these petals is also a mini-flower, usually called a ray floret. These ray florets only have female sexual parts and the petals in each floret fuse, then grow into the long, brightly colored strap that we see as one of the sunflower's petals.

Other flowers

Modifications of the basic floral pattern abound. Snapdragons, peas, and violets add a central axis of symmetry. Mutations in many horticultural chrysanthemums and carnations turn anthers into petals, resulting in flowers with spectacular frills and ruffs of petals around a sexless core. Wind-pollinated flowers like alders and oaks don't bother with petals and have small sepals. Their unisexual flowers either shake pollen from catkins or gather them on female flowers whose stigmas jut into the air. Grasses, too, forgo showy petals and instead turn much of the flower into papery wrappers, parachutes, and hooks.

Floral diversity is a little like music. A few strong themes are elaborated into many gorgeous variations. As well as attracting us with their bursts of color and aroma, they intrigue our minds and emotions with their varied compositions. Sometimes, they puzzle us: Where has *this* flower hidden its anthers? Other times they amaze: Petals can be shaped like *that*? Mostly, they delight us with their interplay of unity and diversity.

Get sexy with an orchid

Everything about orchids is extreme. Their floral displays are over the top. Their relationships with pollinators are intensely codependent and often twisted. Were they humans, therapy would be in order. The parts that actually make pollen and eggs, though, are fused and miniaturized, as are their seeds, the opposite of the boundless extravagance of their sexual displays and relationships.

A little poking in the private parts of the orchids can reveal some of the complexity of their sexuality and relationships with pollinators. If you don't yet have any orchids brightening your windowsill, the easiest to buy and keep alive are moth orchids, *Phalaenopsis.* They are originally from Asia but are now propagated worldwide by horticulturalists. The kinds that you find at supermarkets are mostly hybrids, bred to please human eyes. For this activity, finding a variety with larger blooms is helpful, although with a magnifying lens and a steady hand, you can use small flowers. Choose flowers for this activity that opened in the last week.

Be a bee and look face-to-face at your moth orchid. The two rounded moth "wings" are petals. Behind them is a trio of sepals, flower parts that first acted as green protective wrappers when the flower was in bud, then unclenched and took on the color of the petals as the flower opened.

A column protrudes from the center of the symmetrical display, with a winged lower lip below. The lip is a modified petal, usually curled into a tube that shunts bees into the flower and holds them in just the right position to receive pollen. A small tongue pokes from the back part of the lip. This lure is often covered in lines or speckles, like the visual guides on a runway, and sometimes tinged in yellow, a false signal that pollen is available.

Everything we've seen so far serves to advertise to pollinators, usually bees, and to steer them when they arrive. The flower's goal is to pilot the bees so that they line up directly under the tip of the column where the insect will deliver or receive pollen from hidden sexual parts.

To find these parts, peer closely. At the tip of the protruding column, look for a cap, often the same color as the rest of the column. With a toothpick or the tip of your little finger, prod this cap from below. Imagine that you're a bee jostling along the lip, then backing out. The cap has a little hook on its underside. When you catch it just right, the cap pops off, sometimes springing away, sometimes sticking to the toothpick or finger. Two egg-yolk-yellow pollen sacs sit inside the cap, attached to a sticky flap of tissue. It always astonishes me how little pollen is involved. A big flower for just this? But with the lip guiding bees into position, the orchid has no need of great excess of pollen. These two small sacs will do the job.

If the process feels slightly transgressive, you're right. Your toothpick has initiated sex with an orchid, taking the place of a bee. For me, the weirdness is heightened by the fact that the pollen sacs look embarrassingly like miniature testicles. To do their job, these little sacs have to take wing, stuck to a flying insect, hence their annoying stickiness. You might need a couple of toothpicks to work them out of the cap and onto the tip of one toothpick.

Next, deposit the pollen onto the female part of the flower. When you knocked the cap off, you revealed a depression *under* the tip of the column, often behind a pair of hooks or fangs whose job is to grab the pollen sacs. This is the stigma, often glistening with moisture, where the flower gathers incoming pollen. You'll have the best luck with fertilization if you deliver to the stigma pollen from a different flower, such as another bloom on the same stalk. Depending on how much glue the cap and pollen sacs left on your toothpick, it can be hard to shake the pollen sacs loose. Gently push the sacs up into the depression. Eventually, they'll stick.

Once you've lodged the pollen sacs into the tip of the column, your job as a pollinator is over. Now, patience. Sometimes you can see the stigma close after a day, engulfing the pollen sacs. From here, the fertilization process is invisible. Sometimes the pollen or stigma don't get along or have passed their prime, and nothing happens. If they are friendly and frisky, the pollen sacs open and individual pollen grains grow fine tubes down the column to the ovary, which, at this stage, looks like a boring greenish stalk attaching the flower to the main stem. Eggs nestle inside this slender ovary. If the stigma, column, and eggs decide that the pollen is a good match—a process guided by chemical back-and-forth between female and male cells—sperm cells from pollen and eggs in the ovary fuse. Flowers that successfully breed in this way often drop or shrivel their petals. If you want your blooms to last, keep your sexy toothpick away. Unrequited love stays beautiful longer.

If your matchmaking goes well, after a few weeks the ovary fattens into a fruit shaped like a fat bean pod. After about six months, this fruit will mature into a yellow pod filled with dustlike seeds. Commercial growers germinate and grow these seeds on sterile jelly inside glass flasks. Some will grow your seeds for you—search the internet for "orchid flasking services" or, if you want to try this advanced technique yourself, "DIY orchid flasking." But you can also scatter the seeds among the tangle of orchid stems and moss at the base of the parent plant. Some may germinate. More patience. It takes months or years for these babies to grow into mature flowers.

Even if you don't grow any mature fruits—many fertilization attempts fail—exploring the flowers in this way is fun. Not all sex has to be procreative, even in the floral world. Repeated experiments taught me how wonderfully precise are the architectures of orchid flowers. In particular, I'm astonished at how the female and male parts are fused into a single, compact structure, one that exactly lines up with the ornate guides in the lip. That such a flower should be for sale in the check-out line of a supermarket is an everyday marvel. Getting familiar with the moth orchid also opened my eyes to the diversity of other orchid forms. In the wild, in botanic gardens, and in higher-end flower shops, the range of orchid shapes is breathtaking. What insect fits in *there*? Why is that lip so expansive or frilly? Where are the pollen and stigma hidden on *this* species? A little experimentation with everyday moth orchids at home increases our amazement as we encounter new species.

Moonshine perfume and scented unguent

No, it's not drinkable or alcoholic, but the moonshine perfume is made by at-home distillation. The unguent, a scented oil, requires no heat and is easy to make without any special equipment.

First, distillation. You'll capture the aromatic vapors rising from scented petals. To get raw materials, I surreptitiously clip blooms from the climbing roses that spill onto the sidewalk from some gardens in Atlanta, but scented roses from the store work, too. Be sure to sniff them before buying, as many roses have been bred to look good but smell dull. Put a few handfuls of rose petals, about two cups, in a heavy pan. Only use a pan that has a curved, domed lid, not a flat one. Brush the petals to the edges to clear a central spot on the bottom of the pan. Put a heat-proof glass dish in this spot. This central dish will gather the perfumed water. Add a cup of water to the petals, or more if you have a large pan.

Now, the trick that makes the whole thing work: Put the domed lid on the heavy pan *upside down*. The lid's curve will point downward, with its lowest point over the glass dish. Put some ice into the concave top of the upside-down lid. Turn on the burner under your heavy pan. You want the water to boil, but not too vigorously. Gently does it. The steam will rise inside the pan and then hit the cooled lid. Drip, drip: The condensed steam gathers on the underside of the lid, runs to the middle, and then falls into the waiting glass dish. After about twenty minutes of boiling, the petals and their liquid will be mushy and yellow-brown.

Using oven mitts, carefully take off the inverted lid and lift out the glass dish. The water inside the glass dish will be clear and smell deliciously of roses. The water that the petals boiled in will smell of old lettuce and can be discarded. The aromatic oils from the rose petals wafted upward in the steam, then dripped into the central glass dish. You've captured a dilute essence.

Once it has cooled, you can bottle and enjoy the aromatic rose water at any time. Because this activity involves steam, boiling water, and hot liquids in slippery vessels, put safety first and work with protection for your eyes, skin, and hands. And although rose water can be used in cooking, most commercially grown roses are doused in poisons, so enjoy the aroma of your perfumed rose water but never ingest it.

Next, unguents. These are aromatic ointments and their manufacture is the oldest recorded use of floral perfume. The process is simple and takes a little effort spread over many days. We lay scented petals in fats or soak them in oils, then remove them and replace with fresh petals. Day by day, we build up the scent in the fat or oil. Most aromatic flowers will work. I've used jasmine from our garden and rose petals from the neighborhood. Coconut fat, or "butter," works well and has the advantage that in cooler rooms it solidifies, but on skin it warms to a soothing oil.

Take a lidded glass dish and spoon a centimeter-deep layer of coconut fat onto the bottom. Then poke petals into the top surface, put the lid on, and leave the container at room temperature for twenty-four hours. The messy part comes the next day. I use tweezers to pluck the petals, with a teaspoon on hand in case a petal pulls up too much fat. A swipe with the spoon recovers this precious scented fat, which I smear back into the container. With yesterday's petals gone, add a fresh batch, repeating the daily replacement of petals until your coconut fat has a generous floral aroma. Leaving the petals more than a day or two is a waste of time since most of the aroma flows swiftly to the fat from these thin petals.

After a week or two, you'll have lightly scented moisturizing coconut "butter." You can do a similar process by infusing petals in unscented oils like grapeseed or extra-light olive oil. As with coconut fat, replacing the petals regularly helps to speed up the flow of aroma into the oil and results in a more aromatic end product. I've seen some guides recommend putting the oil on a warm windowsill and leaving it for a week.

That might work for plants like lavender with hefty antibacterial properties, but heat invites rot with jasmine and rose petals.

Perfumery often seems an impossibly complex and subtle art, an idea the perfume industry encourages with images of people in lab coats supplying *maîtres parfumeurs* with hundreds of carefully concocted ingredients. I find that at-home extraction demystifies some of the process, making me feel more connected to flowers and the everyday charms of their aromas. At the same time, homemade extractions have simple aromatic profiles and sometimes fail for mysterious reasons. Too hot? Not enough time? Weakly scented petals? Plain rose water is lovely, but leaves the nose eager for more. The at-home process gives me a greater appreciation for the centuries of art and science that underpin modern perfumery. Home baking is a good analogy. Baking bread at home puts you in touch with yeast and wheat, but also deepens admiration for professionals. Turn your kitchen into a perfumer's lair and see where the experiments lead.

Perfume games in duty-free or department stores

Blooms wait for us at the airport and store, offering an opportunity to play and learn about the cultural symbolism of flowers. Some of these games are solitaire. Others need a partner. All rely on well-stocked rows of test bottles and paper test strips known as fragrance blotters. Spritz some perfume onto a blotter, waft the paper knowingly in the air to dispel the initial burst of alcohol, and then inhale lightly. Wait a minute and repeat to get the deeper notes. If you just huff from the bottle without using a blotter, you'll get only the immediate top notes, missing the aromatic layers underneath.

If you're solo, go to the perfumes with a question. For example, can you find a perfume marketed for men that mentions flowers? Which floral perfumes signal innocence in their packaging and which suggest more sultry inner complexities? Have some flowers been pegged as lively, others as calming, and others as mysteriously exotic? Now, do any of these marketing auras—created by bottle shapes and colors, images on the packaging, banners on shelves or walls—correlate at all with the actual experience of the aroma? When you smell each perfume, where in the body does the feeling manifest? In the nose, obviously, but also perhaps in your neck as you relax, in your heart as you feel a jolt of energy, or as a thrum of aromatic lust flowing down your spine?

If you're with someone else, expand the game into "guess the packaging": Spray four different perfumes onto blotters, then ask your partner to guess which one belongs to which bottle. Sometimes, the connection is easy, a rosy perfume with a slender pink bottle. But others are confounding, such as a depthless, almost silly perfume that emerged from a dark bottle designed to reassure the owner of his square-jawed masculinity. Also with a partner you can play "find the most . . ."—most hideous aroma in the store, or the most sexy or hopeful. Just don't use your own skin and clothes as blotters for more than one or two perfumes: Your seatmate on the flight or the ride home will thank you.

A flower diary

Every day, seek out an encounter with a flower and make note of it. It could be a flower growing wild or in a garden. Or the flower might be in the spice rack (saffron), in the fruit aisle of the supermarket (a fruit is a mature flower), or in a jar of rice (mature grass flowers). Flowers adorn clothing, bloom on magazine pages, glow from paintings, are printed on fabrics, are worked into decorative arts, and find their way into religious and commercial symbols. Whatever the day's flower, jot its name in a notebook. I kept a list for a few months on my phone, making a shortcut to the document from my home screen. But then I bought a small paper diary and found the habit of scribbling names more satisfying, a brief daily ritual of making a material record of floral encounters. Sometimes the flower's name is easy to find, but often we need to poke around in books or online, perhaps using image searches.

I find that this practice of flower seeking turns on an internal "radar." When I'm seeking a flower each day, my senses wake and rove with curiosity. Even when I'm not actively searching, my subconscious turns slightly outward, tuned to blooms. The everyday treasure hunt also taught me how little I know about floral names—often frustratingly so—and this revelation then turned into an admiration for the natural diversity of flowers, the many ways that horticulturalists have reshaped them, and the multifarious ways that flowers are represented in human art. I also sometimes challenge myself to either name the taxonomic family of the flower, get to know who its kin are, or find one interesting fact about the flower's life, such as its relationships with insects or its home of origin.

A daily practice is a delight but also can be hard to sustain, especially in the winter months. When I'm feeling that way, I slow to a weekly flower or take a break—who needs more self-imposed stress?—and instead flip back through the diary, remembering the pleasures of past floral encounters. In a world where so much of our daily news and business centers on humans, a flower diary cracks open the door to news from the more-than-human world. Not only does this chink in the door admit some life-affirming light, the opening reorients our curiosity toward the living world.

Sit with a flower

This is the simplest activity in this short collection of invitations, and the most transformative. Sitting with a flower invites us to deepen attention and expand imagination. In my own life, I was jolted into awareness of the lively interactions of flowers with pollinators by open-ended observation of flowers in a woodland in Tennessee, some of which I recount in *The Forest Unseen*. Lying down with my eye pressed to a hand lens was a dive into the visual beauty of flowers in a way I had never experienced, but it also showed me how each species of flower had an entirely different approach to communicating with insects. A tiny patch of forest floor suddenly felt as densely woven with relationships as a great nineteenth-century novel. Sitting with flowers in containers and small beds outside our kitchen door in Atlanta also taught me how vibrant and action-packed flower lives can be, especially for native plants in the early autumn. Despite the impoverished pollinator community in the city, dozens of bumblebees, wasps, and butterflies swarm asters and goldenrods, each flower attracting a different crew. I also learned how eerily dead some of the flowers I bought from a big-box gardening store were. Witnessing the contrast was soul-chilling in a way that reading no scientific paper has ever been, even though the science is more rigorous than my haphazard experiences.

I'm not alone in this experience of changed "flower consciousness." During one of my conversations with Rebecca McMackin, the ecological horticulturalist who helped to design and run the Brooklyn gardens and parks that I visit for the pansy chapter, she recounted a flower watch. She told me, "I was radicalized by a study at Washington Square Park. I think it was with the Xerces Society [a nonprofit focused on the conservation of invertebrates and their habitats]. They asked us to sit for ten minutes with a flower—goldenrod, *Echinacea*, or a couple of others—and then repeat this sit a few times. We tracked which insects visited. I had been gardening for a while, but that experience made me see . . . I like the term *personhood*." She laughed. "A lot of people disagree with me about that. But flowers are persons. Obviously not human persons. My mind was blown. The level of interaction between the flowers and bees and wasps and others opened up a new world. Gardeners are fortunate to be around flowers, but to have permission to just sit with intention, and then to recognize the desire, communication, agency, and drama. It's so beautiful. It invites people into a secret world."

I suggest three ways of sitting with flowers, each with a slightly different mindset: Sit with a flower and activate your senses, your curiosity, or your imagination. Of course, these are complementary and overlap, but focusing on one can be helpful. Experiment with different flowers: some local to your region, some imported from elsewhere; some legendary for generous nectar (lindens, asters, spring beauties—a brief internet search will turn up some relevant to your home) and others with more stingy and specialized dispositions (orchids, many anemones, and the horticulturally reshaped flowers of pansy, begonia, chrysanthemum, and more).

First, sit and ask your senses to pay attention to the flower. Gaze at the flower as you would a famous painting. Go beyond first impressions into the variations of hue and the details of architecture. When you think you've seen all there is to see, keep looking and wait for other details to emerge. Gently touch and discover the textures of the flower. Stems might have sticky or fluffy hairs to slow unwanted insects. Petals can feel surprisingly warm in the sun and rubbery-cool in the morning. Brush the sexual parts to feel the grip of the stigma and the powdery or granular textures of pollen. Does the pollen stain your fingertips? Bounce your fingers against the flower, using your kinesthetic sense to understand which parts are stout or pliable. Smell, in short little huffs, then in long inhales. What layers can you find in the aroma?

If insects or other visitors arrive, listen for their distinctive sounds. Some pollinators use buzzes to open anthers and release pollen. How does the insect approach the flower? Watch the insects grapple and probe. Do you see pollen caught on the insects' fuzz or the head feathers of the hummingbirds? Watch other humans, too. How do they respond to the flower and to your devotion to its form?

Do all of this without judgment and without too much questioning. Just explore and let the mind be quiet for a while. Or let the mind do its thing but don't follow it. Instead, come repeatedly back to the immediacy of the senses. This embodied exploration lodges the flower into our senses and then our memories. Next time you see a flower of the same kind, you'll experience it more deeply, a plunge into its unique personality.

Another approach is to sit with a flower and to ask questions. Let sensory impressions stir up curiosity. Below, I suggest a swarm of questions, too many for one sit, so pick just one or two at first. Why is this flower open now and not last month? Was the flower open at night? How long will these petals last? How do form and function meld? If you were a bumblebee, a moth, or a short-tongued solitary bee, how might you approach this bloom? Is nectar hidden somewhere here? Where are the ovaries and what fruit will emerge?

Where is this flower's place of origin and who brought it here? What signatures of its original home do you see in its shape and behavior? Has human desire and creativity partly shaped the flower? Who labored to bring this bloom to this place and is their work recognized or honored in any way?

How did the arrangement of the floral parts—petals, sexual bits, and so on—emerge from a tiny bud? Can you find any opening buds that might show you how the flower unfolded itself and swelled from its miniaturized immature form? Is the flower bisexual, and if so, do male and female activate themselves at the same time? Could this flower self-fertilize? Or is the flower expressing just one biological sex?

When you look at the architecture of the bloom, do you sense any convergences or divergences from the shapes of other flowers? What do these commonalities and differences suggest about kinship? What does the aroma remind you of and to whom is it beckoning? How does the flower encourage cooperation and deter nectar and pollen thieves?

What illusions is the flower spinning?

Sitting with active curiosity allows the flower to guide us into discoveries and avenues for exploration. Seek some answers at the library, in the scientific literature, or through web pages focused on floral biology. Like any good teacher, flowers give us a reading list to peruse after the lesson.

Next, sit with a flower and ask your imagination to open to unseen wonders. Human senses can only grasp some of the communication between flowers and their pollinators. The fraught dance between female and male cells within the flower is likewise invisible. But our imaginations can access these stories and we can bring them alive in our minds as we sit. This is "augmented reality," with the augmentation provided by the imaginative capacity that lives inside our minds and emotions. No app needed.

Start by imagining the electrical fields around flowers and pollinators. Flowers are not just visual and aromatic billboards, they are antennae deployed to detect and signal to pollinators using electrical charges. Every flower has a negative charge and its own distinctively shaped electrical field. Bees and hoverflies use special hairs on their bodies to sense the shapes and strengths of these fields. Just as insects learn which colors of flowers have the most nectar, they also learn which shapes of floral electrical fields are most rewarding. For example, daffodils have a ring of electrical charge around the lip of their floral trumpets. Lilies, on the other hand, have strong charges on their prominent anthers. Flying insects carry a positive charge, the opposite of flowers. For both insect and bloom, desire is literally electric. When the insects approach, the electrical field in the flowers immediately changes. Some flowers transmit this change to neighbors, so a cluster of blooms might all sense the arrival of an insect at a single flower. These electrical changes happen over seconds, giving flowers and their pollinators a rapid-fire way of conversing. In places with strong artificial electrical fields, such as under high-voltage power lines, these signals can get scrambled. How will you imagine these halos, exchanges, and distortions? As colors? As sounds?

As you watch the flowers, imagine other unseen connections. Some flowers can hear the buzzing approach of bees and respond by sweetening their nectar. Bees leave aromatic footprints on flowers, tiny dabs of insect perfume that tell others which flowers are rewarding and which are spent. Under the soil, connections between plants and fungi also affect the flowers' conversations with insects. Plants rooted in diverse fungal communities can be more attractive to bees, partly because they make more abundant flowers with richer nectar and pollen. The path of a flying insect among flowers, then, is partly a map of fungus-plant connections buried in the soil.

Inside the flower, pollen grains germinate and are assessed by receptors on the stigma and in the female tissues of the plant. Deep in the ovary, eggs and their companion cells welcome pairs of sperm, newly arrived through slender pollen tubes, and start building the next generation. The ovary senses these unions and, depending on the flower, swells, hardens, or turns into a papery wing. As you sit, let your imagination roam through one or two of these unseen realities.

A flower is not a stationary object, a noun. A flower is not even a verb, a singular process. Rather, a flower is an ingathering of many stories, a beautiful and always changing dance of relationship. Some of these convergences we can sense. Some we have to imagine. When we meet a human friend, we know there is more going on in their head and emotions than they can express in words and facial expressions. So, too, with flowers. Sitting with a bloom and actively imagining all that we cannot directly perceive brings their thrilling complexity to life and also gives us a sense of respect. Wonder and esteem. Surely, the world could use more of both.

Bibliography

T HE STORIES IN THIS BOOK DRAW ON THE PUB-
lished work of thousands of scientists, horticulturalists, and other flower experts. This literature is an extraordinary intellectual ecosystem, a manifestation of human curiosity, wise attention to other species, and care for the world. My book highlights just a few parts of this web of knowledge and work, and I hope that readers will also explore, learn from, and contribute to the richness of this network.

Preface

Hernbrode, Janine, and Peter Boyle. "Flower World Imagery in Petroglyphs: Hints of Hohokam Cosmology on the Landscape." *American Indian Rock Art* 40 (2013): 1077–1092.

Hodgson, Derek, and Paul Pettitt. "The Origins of Iconic Depictions: A Falsifiable Model Derived from the Visual Science of Palaeolithic Cave Art and World Rock Art." *Cambridge Archaeological Journal* 28, no. 4 (2018): 591–612.

Hughes, Becky. "Do Cocktail Glasses Have a Gender? For Some Men, Clearly." *New York Times*, August 25, 2023.

Veth, Peter, Cecilia Myers, Pauline Heaney, and Sven Ouzman. "Plants Before Farming: The Deep History of Plant-Use and Representation in the Rock Art of Australia's Kimberley Region." *Quaternary International* 489 (2018): 26–45.

Walton, Georgina, Jonathan Mitchley, Geraldine Reid, and Sven Batke. "Absence of Botanical European Palaeolithic Cave Art: What Can It Tell Us About Plant Awareness Disparity?" *Plants, People, Planet* 5 (2023): 690–697.

Magnolia

Almeida, Eduardo A. B., Silas Bossert, Bryan N. Danforth, et al. "The Evolutionary History of Bees in Time and Space." *Current Biology* 33, no. 16 (2023): 3409–3422.

Azuma, Hiroshi, Leonard B. Thien, and Shoichi Kawano. "Floral Scents, Leaf Volatiles and Thermogenic Flowers in Magnoliaceae." *Plant Species Biology* 14, no. 2 (1999): 121–127.

Bao, Tong, Bo Wang, Jianguo Li, and David Dilcher. "Pollination of Cretaceous Flowers." *Proceedings of the National Academy of Sciences* 116, no. 49 (2019): 24707–24711.

Barba-Montoya, Jose, Mario Dos Reis, Harald Schneider, Philip C. J. Donoghue, and Ziheng Yang. "Constraining Uncertainty in the Timescale of Angiosperm Evolution and the Veracity of a Cretaceous Terrestrial Revolution." *New Phytologist* 218, no. 2 (2018): 819–834.

Bell, Charles D., Douglas E. Soltis, and Pamela S. Soltis. "The Age and Diversification of the Angiosperms Re-Revisited." *American Journal of Botany* 97, no. 8 (2010): 1296–1303.

Benton, Michael J. "The Origins of Modern Biodiversity on Land." *Philosophical Transactions of the Royal Society B: Biological Sciences* 365, no. 1558 (2010): 3667–3679.

Benton, Michael J., Peter Wilf, and Hervé Sauquet. "The Angiosperm Terrestrial Revolution and the Origins of Modern Biodiversity." *New Phytologist* 233, no. 5 (2022): 2017–2035.

Cardinal, Sophie, and Bryan N. Danforth. "Bees Diversified in the Age of Eudicots." *Proceedings of the Royal Society B: Biological Sciences* 280, no. 1755 (2013): 20122686.

Clarke, John T., Rachel C. M. Warnock, and Philip C. J. Donoghue. "Establishing a Time-Scale for Plant Evolution." *New Phytologist* 192, no. 1 (2011): 266–301.

Coiro, Mario, James A. Doyle, and Jason Hilton. "How Deep Is the Conflict Between Molecular and Fossil Evidence on the Age of Angiosperms?" *New Phytologist* 223, no. 1 (2019): 83–99.

Crane, Peter R., and David L. Dilcher. "Lesqueria: An Early Angiosperm Fruiting Axis from the Mid-Cretaceous." *Annals of the Missouri Botanical Garden* 71 (1984): 384–402.

Darwin, Charles. "Letter 395. To J. D. Hooker." In *More Letters of Charles Darwin. A Record of His Work in a Series of Hitherto Unpublished Letters*, vol. 2. Edited by Francis Darwin and A. C. Seward. London: John Murray, 1903. darwin-online.org.uk.

Dieringer, Gregg, and J. Enrique Espinosa S. "Reproductive Ecology of *Magnolia schiedeana* (Magnoliaceae), a Threatened Cloud Forest Tree Species in Veracruz, Mexico." *Bulletin of the Torrey Botanical Club* 121, no. 2 (1994): 154–159.

Dilcher, David. "Toward a New Synthesis: Major Evolutionary Trends in the Angiosperm Fossil Record." *Proceedings of the National Academy of Sciences* 97, no. 13 (2000): 7030–7036.

Dilcher, David, and Peter R. Crane. "In Pursuit of the First Flower." *Natural History Magazine* 93 (1984): 56–61.

Doyle, James A. "Molecular and Fossil Evidence on the Origin of Angiosperms." *Annual Review of Earth and Planetary Sciences* 40 (2012): 301–326.

Eiseley, Loren. "How Flowers Changed the World." In *The Immense Journey*. Vintage, 1957.

Farrell, Brian D. "'Inordinate Fondness' Explained: Why Are There So Many Beetles?" *Science* 281, no. 5376 (1998): 555–559.

Friis, Else Marie, Kaj Raunsgaard Pedersen, and Peter R. Crane. "Fossil Evidence of Water Lilies (Nymphaeales) in the Early Cretaceous." *Nature* 410, no. 6826 (2001): 357–360.

Friis, Else Marie, Peter R. Crane, and Kaj Raunsgaard Pedersen. *Early Flowers and Angiosperm Evolution*. Cambridge University Press, 2011.

Gitzendanner, Matthew A., Pamela S. Soltis, Gane K.-S. Wong, Brad R. Ruhfel, and Douglas E. Soltis. "Plastid Phylogenomic Analysis of Green Plants: A Billion Years of Evolutionary History." *American Journal of Botany* 105, no. 3 (2018): 291–301.

Gottsberger, Gerhard, Ilse Silberbauer-Gottsberger, Roger S. Seymour, and Stefan Dötterl. "Pollination Ecology of *Magnolia ovata* May Explain the Overall Large Flower Size of the Genus." *Flora* 207, no. 2 (2012): 107–118.

Guo, Xing, Dongming Fang, Sunil Kumar Sahu, et al. "*Chloranthus* Genome Provides Insights into the Early Diversification of Angiosperms." *Nature Communications* 12, no. 1 (2021): 6930.

Hu, Shusheng, David L. Dilcher, David M. Jarzen, and David Winship Taylor. "Early Steps of Angiosperm–Pollinator Coevolution." *Proceedings of the National Academy of Sciences* 105, no. 1 (2008): 240–245.

Jud, Nathan A. "Fossil Evidence for a Herbaceous Diversification of Early Eudicot Angiosperms During the Early Cretaceous." *Proceedings of the Royal Society B: Biological Sciences* 282, no. 1814 (2015): 20151045.

Jud, Nathan A., Michael D. D'Emic, Scott A. Williams, Josh C. Mathews, Katie M. Tremaine, and Janok Bhattacharya. "A New Fossil Assemblage Shows That Large Angiosperm Trees Grew in North America by the Turonian (Late Cretaceous)." *Science Advances* 4, no. 9 (2018): eaar8568.

Kawahara, Akito Y., David Plotkin, Marianne Espeland, et al. "Phylogenomics Reveals the Evolutionary Timing and Pattern of Butterflies and Moths." *Proceedings of the National Academy of Sciences* 116, no. 45 (2019): 22657–22663.

Kawahara, Akito Y., Caroline Storer, Ana Paula S. Carvalho, et al. "A Global Phylogeny of Butterflies Reveals Their Evolutionary History, Ancestral Hosts and Biogeographic Origins." *Nature Ecology & Evolution* 7, no. 6 (2023): 903–913.

Lenton, Timothy M., Tais W. Dahl, Stuart J. Daines, et al. "Earliest Land Plants Created Modern Levels of Atmospheric Oxygen." *Proceedings of the National Academy of Sciences* 113, no. 35 (2016): 9704–9709.

Li, Hong-Tao, Ting-Shuang Yi, Lian-Ming Gao, et al. "Origin of Angiosperms and the Puzzle of the Jurassic Gap." *Nature Plants* 5, no. 5 (2019): 461–470.

Liu, Zhong-Jian, Diying Huang, Chenyang Cai, and Xin Wang. "The Core Eudicot Boom Registered in Myanmar Amber." *Scientific Reports* 8, no. 1 (2018): 16765.

Magallón, Susana, Sandra Gómez-Acevedo, Luna L. Sánchez-Reyes, and Tania Hernández-Hernández. "A Metacalibrated Time-Tree Documents the Early Rise of Flowering Plant Phylogenetic Diversity." *New Phytologist* 207, no. 2 (2015): 437–453.

Manchester, Steven R., David L. Dilcher, Walter S. Judd, Brandon Corder, and James F. Basinger. "Early Eudicot Flower and Fruit: *Dakotanthus* gen. nov. from the Cretaceous Dakota Formation of Kansas and Nebraska, USA." *Acta Palaeobotanica* 58, no. 1 (2018): 27–40.

McKenna, Duane D., Seunggwan Shin, Dirk Ahrens, et al. "The Evolution and Genomic Basis of Beetle Diversity." *Proceedings of the National Academy of Sciences* 116, no. 49 (2019): 24729–24737.

Miller, Rebecca E., Nicole M. Grant, Larry Giles, et al. 'In the Heat of the Night–Alternative Pathway Respiration Drives Thermogenesis in *Philodendron bipinnatifidum*." *New Phytologist* 189, no. 4 (2011): 1013–1026.

Montgomery, Clara, Jozsef Vuts, Christine M. Woodcock, et al. "Bumblebee Electric Charge Stimulates Floral Volatile Emissions in *Petunia integrifolia* but Not in *Antirrhinum majus*." *Science of Nature* 108 (2021): 1–12.

One Thousand Plant Transcriptomes Initiative. "One Thousand Plant Transcriptomes and the Phylogenomics of Green Plants." *Nature* 574, no. 7780 (2019): 679–685.

Paterno, Gustavo Brant, Carina Lima Silveira, Johannes Kollmann, Mark Westoby, and Carlos Roberto Fonseca. "The Maleness of Larger Angiosperm Flowers." *Proceedings of the National Academy of Sciences* 117, no. 20 (2020): 10921–10926.

Pellmyr, Olle, and Leonard B. Thien. "Insect Reproduction and Floral Fragrances: Keys to the Evolution of the Angiosperms?" *Taxon* 35, no. 1 (1986): 76–85.

Ren, Dong, Conrad C. Labandeira, Jorge A. Santiago-Blay, et al. "A Probable Pollination

Mode Before Angiosperms: Eurasian, Long-Proboscid Scorpionflies." *Science* 326, no. 5954 (2009): 840–847.

Romanov, Mikhail S., and David L. Dilcher. "Fruit Structure in Magnoliaceae s.l. and *Archaeanthus* and Their Relationships." *American Journal of Botany* 100, no. 8 (2013): 1494–1508.

Sauquet, Hervé, Maria von Balthazar, Susana Magallón, et al. "The Ancestral Flower of Angiosperms and Its Early Diversification." *Nature Communications* 8, no. 1 (2017): 1–10.

Seymour, Roger S., Ilse Silberbauer-Gottsberger, and Gerhard Gottsberger. "Respiration and Temperature Patterns in Thermogenic Flowers of Magnolia Ovata Under Natural Conditions in Brazil." *Functional Plant Biology* 37, no. 9 (2010): 870–878.

Song, Bo, Lu Sun, Spencer C. H. Barrett, et al. "Global Analysis of Floral Longevity Reveals Latitudinal Gradients and Biotic and Abiotic Correlates." *New Phytologist* 235, no. 5 (2022): 2054–2065.

Sun, Ge, David L. Dilcher, Hongshan Wang, and Zhiduan Chen. "A Eudicot from the Early Cretaceous of China." *Nature* 471, no. 7340 (2011): 625–628.

Taylor, Edith L., Thomas N. Taylor, and Michael Krings. *Paleobotany: The Biology and Evolution of Fossil Plants.* Academic Press, 2009.

Thien, Leonard B. "Floral Biology of Magnolia." *American Journal of Botany* 61, no. 10 (1974): 1037–1045.

Thien, Leonard B., Hiroshi Azuma, and Shoichi Kawano. "New Perspectives on the Pollination Biology of Basal Angiosperms." *International Journal of Plant Sciences* 161, no. S6 (2000): S225–S235.

Veits, Marine, Itzhak Khait, Uri Obolski, et al. "Flowers Respond to Pollinator Sound Within Minutes by Increasing Nectar Sugar Concentration." *Ecology Letters* 22, no. 9 (2019): 1483–1492.

Wang, Wei, David L. Dilcher, Ge Sun, Hong-Shan Wang, and Zhi-Duan Chen. "Accelerated Evolution of Early Angiosperms: Evidence from Ranunculalean Phylogeny by Integrating Living and Fossil Data." *Journal of Systematics and Evolution* 54, no. 4 (2016): 336–341.

Wang, Yongdong, Chengmin Huang, Bainian Sun, Cheng Quan, Jingyu Wu, and Zhicheng Lin. "Paleo-CO2 Variation Trends and the Cretaceous Greenhouse Climate." *Earth-Science Reviews* 129 (2014): 136–147.

Wang, Yu-Bing, Bin-Bin Liu, Ze-Long Nie, et al. "Major Clades and a Revised Classification of *Magnolia* and Magnoliaceae Based on Whole Plastid Genome Sequences via Genome Skimming." *Journal of Systematics and Evolution* 58, no. 5 (2020): 673–695.

Wiegmann, Brian M., Michelle D. Trautwein, Isaac S. Winkler, et al. "Episodic Radiations in the Fly Tree of Life." *Proceedings of the National Academy of Sciences* 108, no. 14 (2011): 5690–5695.

Wu, Shaoyuan, Frank E. Rheindt, Jin Zhang, et al. "Genomes, Fossils, and the Concurrent Rise of Modern Birds and Flowering Plants in the Late Cretaceous." *Proceedings of the National Academy of Sciences* 121, no. 8 (2024): e2319696121.

Xiao, Lifang, Conrad Labandeira, David Dilcher, and Dong Ren. "Florivory of Early Cretaceous Flowers by Functionally Diverse Insects: Implications for Early Angiosperm Pollination." *Proceedings of the Royal Society B* 288, no. 1953 (2021): 20210320.

Zhang, Liangsheng, Fei Chen, Xingtan Zhang, et al. "The Water Lily Genome and the Early Evolution of Flowering Plants." *Nature* 577, no. 7788 (2020): 79–84.

Zuntini, Alexandre R., Tom Carruthers, Olivier Maurin, et al. "Phylogenomics and the Rise of the Angiosperms." *Nature* 629, no. 8013 (2024): 843–850.

Goatsbeard

Barral, Abel, Bernard Gomez, François Fourel, Véronique Daviero-Gomez, and Christophe Lécuyer. "CO2 and Temperature Decoupling at the Million-Year Scale During the Cretaceous Greenhouse." *Scientific Reports* 7, no. 1 (2017): 8310.

Beaulieu, Jeremy M., Ilia J. Leitch, Sunil Patel, et al. "Genome Size Is a Strong Predictor of Cell Size and Stomatal Density in Angiosperms." *New Phytologist* 179, no. 4 (2008): 975–986.

Bomblies, Kirsten. "When Everything Changes at Once: Finding a New Normal After Genome Duplication." *Proceedings of the Royal Society B* 287, no. 1939 (2020): 20202154.

Boyce, C. Kevin, Tim J. Brodribb, Taylor S. Feild, and Maciej A. Zwieniecki. "Angiosperm Leaf Vein Evolution Was Physiologically and Environmentally Transformative." *Proceedings of the Royal Society B: Biological Sciences* 276, no. 1663 (2009): 1771–1776.

Boyce, C. Kevin, Jung-Eun Lee, Taylor S. Feild, Tim J. Brodribb, and Maciej A. Zwieniecki. "Angiosperms Helped Put the Rain in the Rainforests: The Impact of Plant Physiological Evolution on Tropical Biodiversity." *Annals of the Missouri Botanical Garden* 97, no. 4 (2010): 527–540.

Brodribb, Tim J., Taylor S. Feild, and Gregory J. Jordan. "Leaf Maximum Photosynthetic Rate and Venation Are Linked by Hydraulics." *Plant Physiology* 144, no. 4 (2007): 1890–1898.

Brodribb, Tim J., and Taylor S. Feild. "Leaf Hydraulic Evolution Led a Surge in Leaf Photosynthetic Capacity During Early Angiosperm Diversification." *Ecology Letters* 13, no. 2 (2010): 175–183.

Chester, Michael, Malorie J. Lipman, Joseph P. Gallagher, Pamela S. Soltis, and Douglas E. Soltis. "An Assessment of Karyotype Restructuring in the Neoallotetraploid *Tragopogon miscellus* (Asteraceae)." *Chromosome Research* 21 (2013): 75–85.

Clark, James W., and Philip C. J. Donoghue. "Whole-Genome Duplication and Plant Macroevolution." *Trends in Plant Science* 23, no. 10 (2018): 933–945.

De Boer, Hugo Jan, Maarten B. Eppinga, Martin J. Wassen, and Stefan C. Dekker. "A Critical Transition in Leaf Evolution Facilitated the Cretaceous Angiosperm Revolution." *Nature Communications* 3, no. 1 (2012): 1221.

Dehal, Paramvir, and Jeffrey L. Boore. "Two Rounds of Whole Genome Duplication in the Ancestral Vertebrate." *PLoS Biology* 3, no. 10 (2005): e314.

Doyle, Jeff J., and Jeremy E. Coate. "Polyploidy, the Nucleotype, and Novelty: The Impact of Genome Doubling on the Biology of the Cell." *International Journal of Plant Sciences* 180, no. 1 (2019): 1–52.

Dubcovsky, Jorge, and Jan Dvorak. "Genome Plasticity a Key Factor in the Success of Polyploid Wheat Under Domestication." *Science* 316, no. 5833 (2007): 1862–1866.

Feild, Taylor S., Timothy J. Brodribb, Ari Iglesias, et al. "Fossil Evidence for Cretaceous Escalation in Angiosperm Leaf Vein Evolution." *Proceedings of the National Academy of Sciences* 108, no. 20 (2011): 8363–8366.

Fernández, Pol, Rémy Amice, David Bruy, et al. "A 160 Gbp Fork Fern Genome Shatters Size Record for Eukaryotes." *iScience* 27, no. 6 (2024): 109889.

Francis Marion Ownbey Papers, 1934–1974. Manuscripts, Archives, and Special Collections, Washington State University Libraries, Pullman, WA. archiveswest.orbiscascade.org/ark:80444/xv17102.

Franks, Peter J., and David J. Beerling. "Maximum Leaf Conductance Driven by CO2 Effects on Stomatal Size and Density over Geologic Time." *Proceedings of the National Academy of Sciences* 106, no. 25 (2009): 10343–10347.

Huang, Chien-Hsun, Caifei Zhang, Mian Liu, et al. "Multiple Polyploidization Events

Across Asteraceae with Two Nested Events in the Early History Revealed by Nuclear Phylogenomics." *Molecular Biology and Evolution* 33, no. 11 (2016): 2820–2835.

Jepsen, Glenn L., Ernst Mayr, George Gaylord Simpson, eds. *Genetics, Paleontology, and Evolution*. Princeton University Press, 1949. wellcomecollection.org/works/gfkjjyc5.

Jiao, Yuannian, Jingping Li, Haibao Tang, and Andrew H. Paterson. "Integrated Syntenic and Phylogenomic Analyses Reveal an Ancient Genome Duplication in Monocots." *Plant Cell* 26, no. 7 (2014): 2792–2802.

Jiao, Yuannian, Norman J. Wickett, Saravanaraj Ayyampalayam, et al. "Ancestral Polyploidy in Seed Plants and Angiosperms." *Nature* 473, no. 7345 (2011): 97–100.

Jordon-Thaden, Ingrid E., Jonathan P. Spoelhof, Lyderson Facio Viccini, et al. "Phenotypic Trait Variation in the North American *Tragopogon* Allopolyploid Complex." *American Journal of Botany* 110, no. 7 (2023): e16189.

Landis, Jacob B., Douglas E. Soltis, Zheng Li, et al. "Impact of Whole-Genome Duplication Events on Diversification Rates in Angiosperms." *American Journal of Botany* 105, no. 3 (2018): 348–363.

Li, Xiaoyu, Weina Si, QianQian Qin, Hao Wu, and Haiyang Jiang. "Deciphering Evolutionary Dynamics of *SWEET* Genes in Diverse Plant Lineages." *Scientific Reports* 8, no. 1 (2018): 13440.

Lim, K. Yoong, Douglas E. Soltis, Pamela S. Soltis, et al. "Rapid Chromosome Evolution in Recently Formed Polyploids in *Tragopogon* (Asteraceae)." *PLoS One* 3, no. 10 (2008): e3353.

Ownbey, Marion. "Natural Hybridization and Amphiploidy in the Genus *Tragopogon*." *American Journal of Botany* 37, no. 7 (1950): 487–499.

Pires, J. Chris, K. Yoong Lim, Ales Kovarík, et al. "Molecular Cytogenetic Analysis of Recently Evolved *Tragopogon* (Asteraceae) Allopolyploids Reveal a Karyotype That Is Additive of the Diploid Progenitors." *American Journal of Botany* 91, no. 7 (2004): 1022–1035.

Sankoff, David, and Chunfang Zheng. "Whole Genome Duplication in Plants: Implications for Evolutionary Analysis." *Comparative Genomics: Methods and Protocols* (2018): 291–315.

Simonin, Kevin A., and Adam B. Roddy. "Genome Downsizing, Physiological Novelty, and the Global Dominance of Flowering Plants." *PLoS Biology* 16, no. 1 (2018): e2003706.

Soltis, Douglas E., Pamela S. Soltis, J. Chris Pires, Ales Kovarik, Jennifer A. Tate, and Evgeny Mavrodiev. "Recent and Recurrent Polyploidy in *Tragopogon* (Asteraceae): Cytogenetic, Genomic and Genetic Comparisons." *Biological Journal of the Linnean Society* 82, no. 4 (2004): 485–501.

Soltis, Douglas E., Richard J. A. Buggs, W. Brad Barbazuk, et al. "The Early Stages of Polyploidy: Rapid and Repeated Evolution in *Tragopogon*." In *Polyploidy and Genome Evolution*, edited by Pamela S. Soltis and Douglas E. Soltis. Springer, 2012.

Soltis, Douglas E., Evgeny V. Mavrodiev, Matthew A. Gitzendanner, Yuri E. Alexeev, Grant T. Godden, and Pamela S. Soltis. "*Tragopogon dubius*: Multiple Introductions to North America and the Formation of the New World Tetraploids." *Taxon* 71, no. 6 (2022): 1287–1298.

Soltis, Douglas E., Evgeny V. Mavrodiev, Vladimir Brukhin, et al. "*Tragopogon pratensis*: Multiple Introductions to North America, Circumscription, and the Formation of the Allotetraploid *T. miscellus*." *Taxon* 72, no. 4 (2023): 848–861.

Soltis, Douglas E., Jennifer A. Tate, Pamela S. Soltis, and V. Vaughan Symonds. "Parental Genomic Compatibility Model: Only Certain Diploid Genotype Combinations Form Allopolyploids." *The Nucleus* 66, no. 3 (2023): 371–378.

Soltis, Pamela S., and Douglas E. Soltis. "Ancient WGD Events as Drivers of Key Innovations in Angiosperms." *Current Opinion in Plant Biology* 30 (2016): 159–165.

Spoelhof, Jonathan P., Douglas E. Soltis, and Pamela S. Soltis. "Polyploidy and Mutation in *Arabidopsis*." *Evolution* 75, no. 9 (2021): 2299–2308.

Stravinsky, Igor, and Robert Craft. *Conversations with Igor Stravinsky*. Doubleday, 1959.

Tang, Dié, Yuxin Jia, Jinzhe Zhang, et al. "Genome Evolution and Diversity of Wild and Cultivated Potatoes." *Nature* 606, no. 7914 (2022): 535–541.

Van de Peer, Yves, Tia-Lynn Ashman, Pamela S. Soltis, and Douglas E. Soltis. "Polyploidy: An Evolutionary and Ecological Force in Stressful Times." *Plant Cell* 33, no. 1 (2021): 11–26.

Vanneste, Kevin, Guy Baele, Steven Maere, and Yves Van de Peer. "Analysis of 41 Plant Genomes Supports a Wave of Successful Genome Duplications in Association with the Cretaceous–Paleogene Boundary." *Genome Research* 24, no. 8 (2014): 1334–1347.

Orchid

Ackerman, James D., Ryan D. Phillips, Raymond L. Tremblay, et al. "Beyond the Various Contrivances by Which Orchids Are Pollinated: Global Patterns in Orchid Pollination Biology." *Botanical Journal of the Linnean Society* 202, no. 3 (2023): 295–324.

Baguette, Michel, Joris A. M. Bertrand, Virginie M. Stevens, and Bertrand Schatz. "Why Are There So Many Bee-Orchid Species? Adaptive Radiation by Intra-Specific Competition for Mnesic Pollinators." *Biological Reviews* 95, no. 6 (2020): 1630–1663.

Barriault, I., Denis Barabé, L. Cloutier, and Marc Gibernau. "Pollination Ecology and Reproductive Success in Jack-in-the-Pulpit (*Arisaema triphyllum*) in Québec (Canada)." *Plant Biology* 12, no. 1 (2010): 161–171.

Boberg, Elin, Ronny Alexandersson, Magdalena Jonsson, Johanne Maad, Jon Ågren, and L. Anders Nilsson. "Pollinator Shifts and the Evolution of Spur Length in the Moth-Pollinated Orchid *Platanthera bifolia*." *Annals of Botany* 113, no. 2 (2014): 267–275.

Breitkopf, Hendrik, Renske E. Onstein, Donata Cafasso, Philipp M. Schlüter, and Salvatore Cozzolino. "Multiple Shifts to Different Pollinators Fuelled Rapid Diversification in Sexually Deceptive *Ophrys* Orchids." *New Phytologist* 207, no. 2 (2015): 377–389.

Brunton-Martin, Amy L., James C. O'Hanlon, and Anne C. Gaskett. "Orchid Sexual Deceit Affects Pollinator Sperm Transfer." *Functional Ecology* 34, no. 7 (2020): 1336–1344.

Brunton-Martin, Amy L., James C. O'Hanlon, and Anne C. Gaskett. "Are Some Species 'Robust' to Exploitation? Explaining Persistence in Deceptive Relationships." *Evolutionary Ecology* 36, no. 3 (2022): 321–339.

Buggs, Richard J. A. "The Deepening of Darwin's Abominable Mystery." *Nature Ecology & Evolution* 1, no. 6 (2017): 0169.

Callan, E. McC. "Nesting Behavior and Prey of *Argogorytes* Ashmead (Hymenoptera: Sphecidae)." *Journal of the Washington Academy of Sciences* 70, no. 4 (1980): 160–165.

Cameron, Duncan D., Jonathan R. Leake, and David J. Read. "Mutualistic Mycorrhiza in Orchids: Evidence from Plant–Fungus Carbon and Nitrogen Transfers in the Green-Leaved Terrestrial Orchid *Goodyera repens*." *New Phytologist* 171, no. 2 (2006): 405–416.

Cameron, Duncan D., Irene Johnson, David J. Read, and Jonathan R. Leake. "Giving and Receiving: Measuring the Carbon Cost of Mycorrhizas in the Green Orchid, *Goodyera repens*." *New Phytologist* 180, no. 1 (2008): 176–184.

Claessens, Jean, and Jacques Kleynen. "Investigations on the Autogamy in *Ophrys apifera* Hudson." *Jahresberichte des Naturwissenschlaftlichen Vereins Wuppertal* 55 (2002): 62–77.

Darwin, Charles. *The Various Contrivances by Which Orchids Are Fertilised by Insects*. 2nd ed.

London: John Murray, 1877. Quoted from the Darwin Online archive, darwin-online .org.uk.

Davies, Kevin L., Malgorzata Stpiczyńska, and Magdalena Kamińska. "Dual Deceit in Pseudopollen-Producing *Maxillaria s.s.* (Orchidaceae: Maxillariinae)." *Botanical Journal of the Linnean Society* 173, no. 4 (2013): 744–763.

Dearnaley, John D. W., Florent Martos, and M.-A. Selosse. "Orchid Mycorrhizas: Molecular Ecology, Physiology, Evolution and Conservation Aspects." In *Fungal Associations*, edited by Bertold Hock. Springer, 2012.

Else, G. R. "*Eucera longicornis* (Linnaeus, 1758)." Bees, Wasps & Ants Recording Society, 2012. bwars.com/bee/apidae/eucera-longicornis.

Gervasi, Daniel D. L., and Florian P. Schiestl. "Real-Time Divergent Evolution in Plants Driven by Pollinators." *Nature Communications* 8, no. 1 (2017): 14691.

Huang, Chao-Li, Feng-Yin Jian, Hao-Jen Huang, et al. "Deciphering Mycorrhizal Fungi in Cultivated *Phalaenopsis* Microbiome with Next-Generation Sequencing of Multiple Barcodes." *Fungal Diversity* 66 (2014): 77–88.

Hughes, Nicole M., Michaela K. Connors, Mary H. Grace, Mary Ann Lila, Brooke N. Willans, and Andrew J. Wommack. "The Same Anthocyanins Served Four Different Ways: Insights into Anthocyanin Structure-Function Relationships from the Wintergreen Orchid, *Tipularia discolor.*" *Plant Science* 303 (2021): 110793.

Hughes, Nicole M., and Simcha Lev-Yadun. "Why Do Some Plants Have Leaves with Red or Purple Undersides?" *Environmental and Experimental Botany* 205 (2023): 105126.

Inda, Luis A., Manuel Pimentel, and Mark W. Chase. "Phylogenetics of Tribe Orchideae (Orchidaceae: Orchidoideae) Based on Combined DNA Matrices: Inferences Regarding Timing of Diversification and Evolution of Pollination Syndromes." *Annals of Botany* 110, no. 1 (2012): 71–90.

Johnson, Steven D., and Florian P. Schiestl. *Floral Mimicry.* Oxford University Press, 2016.

Kevan, Peter G., D. Eisikowitch, John D. Ambrose, and James R. Kemp. "Cryptic Dioecy and Insect Pollination in *Rosa setigera* Michx. (Rosaceae), a Rare Plant of Carolinian Canada." *Biological Journal of the Linnean Society* 40, no. 3 (1990): 229–243.

Lunau, Klaus. "The Ecology and Evolution of Visual Pollen Signals." *Plant Systematics and Evolution* 222 (2000): 89–111.

McCormick, Melissa, Robert Burnett, and Dennis Whigham. "Protocorm-Supporting Fungi Are Retained in Roots of Mature *Tipularia discolor* Orchids as Mycorrhizal Fungal Diversity Increases." *Plants* 10, no. 6 (2021): 1251.

McKendrick, S. L., J. R. Leake, D. Lee Taylor, and D. J. Read. "Symbiotic Germination and Development of the Myco-Heterotrophic Orchid *Neottia nidus-avis* in Nature and Its Requirement for Locally Distributed *Sebacina* spp." *New Phytologist* 154, no. 1 (2002): 233–247.

Paulus, Hannes F. "Pollinators as Isolation Mechanisms: Field Observations and Field Experiments Regarding Specificity of Pollinator Attraction in the Genus *Ophrys* (Orchidaceae und Insecta, Hymenoptera, Apoidea)." *Entomologia Generalis* 37, no. 3–4 (2018): 261–316.

Pérez-Escobar, O. A., D. Bogarín, N. A. Przelomska, et al. "The Origin and Speciation of Orchids." *New Phytologist* 242, no. 2 (2024): 700–716.

Ramos, Sergio E., and Florian P. Schiestl. "Rapid Plant Evolution Driven by the Interaction of Pollination and Herbivory." *Science* 364, no. 6436 (2019): 193–196.

Rasmussen, Hanne N. "Recent Developments in the Study of Orchid Mycorrhiza." *Plant and Soil* 244 (2002): 149–163.

Rasmussen, Hanne N., and Dennis F. Whigham. "Importance of Woody Debris in Seed Germination of *Tipularia discolor* (Orchidaceae)." *American Journal of Botany* 85, no. 6 (1998): 829–834.

Schatz, Bertrand, David Genoud, Pascal Escudié, Philippe Geniez, Karl Günter Wünsch, and Nina Joffard. "Is *Ophrys* Pollination More Opportunistic Than Previously Thought? Insights from Different Field Methods of Pollinator Observation." *Botany Letters* 168, no. 3 (2021): 333–347.

Schiebold, Julienne M.-I., Martin I. Bidartondo, Florian Lenhard, Andreas Makiola, and Gerhard Gebauer. "Exploiting Mycorrhizas in Broad Daylight: Partial Mycoheterotrophy Is a Common Nutritional Strategy in Meadow Orchids." *Journal of Ecology* 106, no. 1 (2018): 168–178.

Schiestl, Florian P., Manfred Ayasse, Hannes F. Paulus, et al. "Orchid Pollination by Sexual Swindle." *Nature* 399, no. 6735 (1999): 421–422.

Schlüter, P., A. Russo, M. Alessandrini, et al. "The Genome of the Early Spider-Orchid *Ophrys sphegodes* Provides Insights into Sexual Deception and Adaptation to Pollinators." 2023. Preprint online, researchsquare.com/article/rs-3463148/v1.

Scopece, Giovanni, Alex Widmer, and Salvatore Cozzolino. "Evolution of Postzygotic Reproductive Isolation in a Guild of Deceptive Orchids." *American Naturalist* 171, no. 3 (2008): 315–326.

Shrestha, Mani, A. G. Dyer, Alan Dorin, Z.-X. Ren, and M. Burd. "Rewardlessness in Orchids: How Frequent and How Rewardless?" *Plant Biology* 22, no. 4 (2020): 555–561.

Snow, Allison A., and Dennis F. Whigham. "Costs of Flower and Fruit Production in *Tipularia discolor* (Orchidaceae)." *Ecology* 70, no. 5 (1989): 1286–1293.

Stevenson, Philip C. "For Antagonists and Mutualists: The Paradox of Insect Toxic Secondary Metabolites in Nectar and Pollen." *Phytochemistry Reviews* 19, no. 3 (2020): 603–614.

Stoutamire, Warren. "Pollination of *Tipularia discolor*, an Orchid with Modified Symmetry." *American Orchid Society Bulletin* 47, no. 5 (1978): 413–415.

Stuppy, Wolfgang. "Orchid Seeds: Nature's Tiny Treasures." Royal Botanic Gardens, Kew. www.kew.org/read-and-watch/orchid-seeds-natures-tiny-treasures.

Thatcher, Margaret. "Interview for *Woman's Own*, 23 September 1987." Margaret Thatcher Foundation. www.margaretthatcher.org/document/106689.

Thompson, Jamie B., Katie E. Davis, Harry O. Dodd, Matthew A. Wills, and Nicholas K. Priest. "Speciation Across the Earth Driven by Global Cooling in Terrestrial Orchids." *Proceedings of the National Academy of Sciences* 120, no. 29 (2023): e2102408120.

"*Tipularia discolor* (Pursh) Nutt. Cranefly Orchid." North American Orchid Conservation Center. goorchids.northamericanorchidcenter.org/species/tipularia/discolor.

Van der Niet, Timotheüs, and Steven D. Johnson. "Phylogenetic Evidence for Pollinator-Driven Diversification of Angiosperms." *Trends in Ecology & Evolution* 27, no. 6 (2012): 353–361.

Van der Niet, Timotheüs, Rod Peakall, and Steven D. Johnson. "Pollinator-Driven Ecological Speciation in Plants: New Evidence and Future Perspectives." *Annals of Botany* 113, no. 2 (2014): 199–212.

Vogel, Stefan, and Jochen Martens. "A Survey of the Function of the Lethal Kettle Traps of *Arisaema* (Araceae), with Records of Pollinating Fungus Gnats from Nepal." *Botanical Journal of the Linnean Society* 133, no. 1 (2000): 61–100.

Waterman, Richard J., and Martin I. Bidartondo. "Deception Above, Deception Below: Linking Pollination and Mycorrhizal Biology of Orchids." *Journal of Experimental Botany* 59, no. 5 (2008): 1085–1096.

Whigham, Dennis F., and Margaret McWethy. "Studies on the Pollination Ecology of *Tipularia discolor* (Orchidaceae)." *American Journal of Botany* 67, no. 4 (1980): 550–555.

Wright, G. A., D. D. Baker, M. J. Palmer, et al. "Caffeine in Floral Nectar Enhances a Pollinator's Memory of Reward." *Science* 339, no. 6124 (2013): 1202–1204.

Xu, Shuqing, Philipp M. Schlüter, Giovanni Scopece, et al. "Floral Isolation Is the Main Reproductive Barrier Among Closely Related Sexually Deceptive Orchids." *Evolution* 65, no. 9 (2011): 2606–2620.

Xu, Shuqing, Philipp M. Schlüter, Ueli Grossniklaus, and Florian P. Schiestl. "The Genetic Basis of Pollinator Adaptation in a Sexually Deceptive Orchid." *PLoS Genetics* 8, no. 8 (2012): e1002889.

Zhang, Guojin, Yi Hu, Ming-Zhong Huang, et al. "Comprehensive Phylogenetic Analyses of Orchidaceae Using Nuclear Genes and Evolutionary Insights into Epiphytism." *Journal of Integrative Plant Biology* 65, no. 5 (2023): 1204–1225.

Grass

Ad Council. "Smokey Bear Celebrates 75th Birthday with Celebrity Friends in Innovative New Animated Emoji Campaign." News release, April 4, 2019. prnewswire.com/news-releases/smokey-bear-celebrates-75th-birthday-with-celebrity-friends-in-innovative-new-animated-emoji-campaign-300824724.html.

Andrae, J. W., F. A. McInerney, P. J. Polissar, et al. "Initial Expansion of C4 Vegetation in Australia During the Late Pliocene." *Geophysical Research Letters* 45, no. 10 (2018): 4831–4840.

Badgley, Catherine, John C. Barry, Michèle E. Morgan, et al. "Ecological Changes in Miocene Mammalian Record Show Impact of Prolonged Climatic Forcing." *Proceedings of the National Academy of Sciences* 105, no. 34 (2008): 12145–12149.

Bai, Yongfei, and M. Francesca Cotrufo. "Grassland Soil Carbon Sequestration: Current Understanding, Challenges, and Solutions." *Science* 377, no. 6606 (2022): 603–608.

Bar-On, Yinon M., and Ron Milo. "The Global Mass and Average Rate of Rubisco." *Proceedings of the National Academy of Sciences* 116, no. 10 (2019): 4738–4743.

Bolognini, Davide, Alma Halgren, Runyang Nicolas Lou, et al. "Recurrent Evolution and Selection Shape Structural Diversity at the Amylase Locus." *Nature* 634, no. 8034 (2024): 617–625.

Bond, William J. "Fires in the Cenozoic: A Late Flowering of Flammable Ecosystems." *Frontiers in Plant Science* 5 (2015): 117055.

Cantalapiedra, Juan L., Richard G. FitzJohn, Tyler S. Kuhn, et al. "Dietary Innovations Spurred the Diversification of Ruminants During the Caenozoic." *Proceedings of the Royal Society B: Biological Sciences* 281, no. 1776 (2014): 20132746.

Cardona, Tanai. "Early Archean Origin of Heterodimeric Photosystem I." *Heliyon* 4, no. 3 (2018).

Cenozoic CO2 Proxy Integration Project (CenCO2PIP) Consortium. "Toward a Cenozoic History of Atmospheric CO2." *Science* 382, no. 6675 (2023): eadi5177.

Charles-Dominique, Tristan, T. Jonathan Davies, Gareth P. Hempson, et al. "Spiny Plants, Mammal Browsers, and the Origin of African Savannas." *Proceedings of the National Academy of Sciences* 113, no. 38 (2016): E5572–E5579.

Christin, Pascal-Antoine, Guillaume Besnard, Emanuela Samaritani, et al. "Oligocene CO2 Decline Promoted C4 Photosynthesis in Grasses." *Current Biology* 18, no. 1 (2008): 37–43.

Craven, Dylan, Forest Isbell, Pete Manning, et al. "Plant Diversity Effects on Grassland Productivity Are Robust to Both Nutrient Enrichment and Drought." *Philosophical Transactions of the Royal Society B: Biological Sciences* 371, no. 1694 (2016): 20150277.

Davis, Katie E., Adam T. Bakewell, Jon Hill, Hojun Song, and Peter Mayhew. "Global Cooling & the Rise of Modern Grasslands: Revealing Cause & Effect of Environmental Change on Insect Diversification Dynamics." bioRxiv [preprint], 2018. doi.org /10.1101/392712.

Edwards, Erika J., Colin P. Osborne, Caroline A. E. Strömberg, et al. "The Origins of C4 Grasslands: Integrating Evolutionary and Ecosystem Science." *Science* 328, no. 5978 (2010): 587–591.

Erb, Tobias J., and Jan Zarzycki. "A Short History of RubisCO: The Rise and Fall (?) of Nature's Predominant CO2 Fixing Enzyme." *Current Opinion in Biotechnology* 49 (2018): 100–107.

Flint-Garcia, Sherry A. "Genetics and Consequences of Crop Domestication." *Journal of Agricultural and Food Chemistry* 61, no. 35 (2013): 8267–8276.

Folk, Ryan A., Carolina M. Siniscalchi, and Douglas E. Soltis. "Angiosperms at the Edge: Extremity, Diversity, and Phylogeny." *Plant, Cell & Environment* 43, no. 12 (2020): 287–2893.

Food and Agriculture Organization of the United Nations. *Agricultural Production Statistics 2000–2022.* FAOSTAT Analytical Briefs, no. 79. Rome, 2023. doi.org/10.4060/cc 9205en.

Friedman, William E. "The Evolution of Embryogeny in Seed Plants and the Developmental Origin and Early History of Endosperm." *American Journal of Botany* 81, no. 11 (1994): 1468–1486.

Gallaher, Timothy J., Paul M. Peterson, Robert J. Soreng, et al. "Grasses Through Space and Time: An Overview of the Biogeographical and Macroevolutionary History of Poaceae." *Journal of Systematics and Evolution* 60, no. 3 (2022): 522–569.

Glasspool, Ian J., Andrew C. Scott, David Waltham, Natalia Pronina, and Longyi Shao. "The Impact of Fire on the Late Paleozoic Earth System." *Frontiers in Plant Science* 6 (2015): 756.

Grange, Guylain, John A. Finn, and Caroline Brophy. "Plant Diversity Enhanced Yield and Mitigated Drought Impacts in Intensively Managed Grassland Communities." *Journal of Applied Ecology* 58, no. 9 (2021): 1864–1875.

Huang, Weichen, Lin Zhang, J. Travis Columbus, et al. "A Well-Supported Nuclear Phylogeny of Poaceae and Implications for the Evolution of C4 Photosynthesis." *Molecular Plant* 15, no. 4 (2022): 755–777.

Jernvall, Jukka, and Mikael Fortelius. "Common Mammals Drive the Evolutionary Increase of Hypsodonty in the Neogene." *Nature* 417, no. 6888 (2002): 538–540.

Karp, Allison T., J. Tyler Faith, Jennifer R. Marlon, and A. Carla Staver. "Global Response of Fire Activity to Late Quaternary Grazer Extinctions." *Science* 374, no. 6571 (2021): 1145–1148.

Kennett, Douglas J., Keith M. Prufer, Brendan J. Culleton, et al. "Early Isotopic Evidence for Maize as a Staple Grain in the Americas." *Science Advances* 6, no. 23 (2020): eaba3245.

Keyser, Patrick D., Dennis W. Hancock, Landon Marks, and Leanne Dillard. *Establishing Native Grass Forages in the Southeast—PB 1873.* University of Tennessee Extension, 2015.

Kumar, Santosh, Milan Soukup, and Rivka Elbaum. "Silicification in Grasses: Variation Between Different Cell Types." *Frontiers in Plant Science* 8 (2017): 438.

Kunimatsu, Yutaka, Masato Nakatsukasa, Yoshihiro Sawada, et al. "A New Late Miocene Great Ape from Kenya and Its Implications for the Origins of African Great Apes and Humans." *Proceedings of the National Academy of Sciences* 104, no. 49 (2007): 19220–19225.

Larrasoaña, Juan C., Andrew P. Roberts, and Eelco J. Rohling. "Dynamics of Green Sahara Periods and Their Role in Hominin Evolution." *PLoS One* 8, no. 10 (2013): e76514.

Lehmer, O. R., D. C. Catling, R. Buick, D. E. Brownlee, and S. Newport. "Atmospheric

CO2 Levels from 2.7 Billion Years Ago Inferred from Micrometeorite Oxidation." *Science Advances* 6, no. 4 (2020): eaay4644.

Linder, H. Peter, Caroline E. R. Lehmann, Sally Archibald, Colin P. Osborne, and David M. Richardson. "Global Grass (Poaceae) Success Underpinned by Traits Facilitating Colonization, Persistence and Habitat Transformation." *Biological Reviews* 93, no. 2 (2018): 1125–1144.

Lord, Janice M., and Mark Westoby. "Accessory Costs of Seed Production and the Evolution of Angiosperms." *Evolution* 66, no. 1 (2012): 200–210.

Maeda, Hiroshi A., and Alisdair R. Fernie. "Evolutionary History of Plant Metabolism." *Annual Review of Plant Biology* 72 (2021): 185–216.

Nerlekar, Ashish N., and Joseph W. Veldman. "High Plant Diversity and Slow Assembly of Old-Growth Grasslands." *Proceedings of the National Academy of Sciences* 117, no. 31 (2020): 18550–18556.

Noss, Reed F. *Forgotten Grasslands of the South: Natural History and Conservation.* Island Press, 2012.

Noss, Reed F., Jennifer M. Cartwright, Dwayne Estes, et al. *Science Needs of Southeastern Grassland Species of Conservation Concern: A Framework for Species Status Assessments.* Open-File Report 2021-1047. US Geological Survey, 2021.

Nunes, Tiago D. G., Dan Zhang, and Michael T. Raissig. "Form, Development and Function of Grass Stomata." *Plant Journal* 101, no. 4 (2020): 780–799.

Oliver, Thomas, Tom D. Kim, Joko P. Trinugroho, et al. "The Evolution and Evolvability of Photosystem II." *Annual Review of Plant Biology* 74 (2023): 225–257.

O'Mara, Frank P. "The Role of Grasslands in Food Security and Climate Change." *Annals of Botany* 110, no. 6 (2012): 1263–1270.

Osborne, Colin P. "Atmosphere, Ecology and Evolution: What Drove the Miocene Expansion of C4 Grasslands?" *Journal of Ecology* 96, no. 1 (2008): 35–45.

Pardo, Jeremy, and Robert VanBuren. "Evolutionary Innovations Driving Abiotic Stress Tolerance in C4 Grasses and Cereals." *Plant Cell* 33, no. 11 (2021): 3391–3401.

Patterson, David B., David R. Braun, Kayla Allen, et al. "Comparative Isotopic Evidence from East Turkana Supports a Dietary Shift Within the Genus *Homo*." *Nature Ecology & Evolution* 3, no. 7 (2019): 1048–1056.

Prasad, V., C. A. E. Strömberg, A. D. Leaché, et al. "Late Cretaceous Origin of the Rice Tribe Provides Evidence for Early Diversification in Poaceae." *Nature Communications* 2, no. 1 (2011): 480.

Quinn, Rhonda L., and Christopher J. Lepre. "Contracting Eastern African C4 Grasslands During the Extinction of *Paranthropus boisei*." *Scientific Reports* 11, no. 1 (2021): 7164.

Robinson, Joshua R., John Rowan, Christopher J. Campisano, Jonathan G. Wynn, and Kaye E. Reed. "Late Pliocene Environmental Change During the Transition from *Australopithecus* to *Homo*." *Nature Ecology & Evolution* 1, no. 6 (2017): 0159.

Ruiz-Giralt, Abel, Laurie Nixon-Darcus, A. Catherine D'Andrea, Yemane Meresa, Stefano Biagetti, and Carla Lancelotti. "On the Verge of Domestication: Early Use of C4 Plants in the Horn of Africa." *Proceedings of the National Academy of Sciences* 120, no. 27 (2023): e2300166120.

Sabelli, Paolo A., and Brian A. Larkins. "The Development of Endosperm in Grasses." *Plant Physiology* 149, no. 1 (2009): 14–26.

Sage, Rowan F., Pascal-Antoine Christin, and Erika J. Edwards. "The C4 Plant Lineages of Planet Earth." *Journal of Experimental Botany* 62, no. 9 (2011): 3155–3169.

Schoeninger, Margaret J. "Stable Isotope Analyses and the Evolution of Human Diets." *Annual Review of Anthropology* 43 (2014): 413–430.

Schubert, Marian, Lars Grønvold, Simen R. Sandve, Torgeir R Hvidsten, and Siri Fjellheim. "Evolution of Cold Acclimation and Its Role in Niche Transition in the Temperate Grass Subfamily Pooideae." *Plant Physiology* 180, no. 1 (2019): 404–419.

Ségalen, Loïc, Julia A. Lee-Thorp, and Thure Cerling. "Timing of C4 Grass Expansion Across Sub-Saharan Africa." *Journal of Human Evolution* 53, no. 5 (2007): 549–559.

Sheldon, Nathan D. "Precambrian Paleosols and Atmospheric CO2 Levels." *Precambrian Research* 147, no. 1–2 (2006): 148–155.

Strömberg, Caroline A. E. "Evolution of Hypsodonty in Equids: Testing a Hypothesis of Adaptation." *Paleobiology* 32, no. 2 (2006): 236–258.

Strömberg, Caroline A. E. "Evolution of Grasses and Grassland Ecosystems." *Annual Review of Earth and Planetary Sciences* 39 (2011): 517–544.

Tipple, Brett J., and Mark Pagani. "The Early Origins of Terrestrial C4 Photosynthesis." *Annual Review of Earth Planetary Science* 35 (2007): 435–461.

Ungar, Peter S., and Matt Sponheimer. "The Diets of Early Hominins." *Science* 334, no. 6053 (2011): 190–193.

United States Patent and Trademark Office. "Burner Bob." Trademark registration number 5514683, filed January 12, 2017, and registered July 10, 2018.

Uno, Kevin T., Thure E. Cerling, John M. Harris, et al. "Late Miocene to Pliocene Carbon Isotope Record of Differential Diet Change Among East African Herbivores." *Proceedings of the National Academy of Sciences* 108, no. 16 (2011): 6509–6514.

Wadley, Lyn, Lucinda Backwell, Francesco d'Errico, and Christine Sievers. "Cooked Starchy Rhizomes in Africa 170 Thousand Years Ago." *Science* 367, no. 6473 (2020): 87–91.

Wynn, Jonathan G., Zeresenay Alemseged, René Bobe, Frederick E. Grine, Enquye W. Negash, and Matt Sponheimer. "Isotopic Evidence for the Timing of the Dietary Shift Toward C4 Foods in Eastern African *Paranthropus.*" *Proceedings of the National Academy of Sciences* 117, no. 36 (2020): 21978–21984.

Yokota, Akiho. "Revisiting RuBisCO." *Bioscience, Biotechnology, and Biochemistry* 81, no. 11 (2017): 2039–2049.

Seagrass

Ackerman, Josef Daniel. "Submarine Pollination in the Marine Angiosperm *Zostera marina* (Zosteraceae). I. The Influence of Floral Morphology on Fluid Flow." *American Journal of Botany* 84, no. 8 (1997): 1099–1109.

Amos, Ilona. "Seagrass Planting Set to Begin in Firth of Forth as Major Restoration Project Gathers Pace." *The Scotsman*, February 2, 2023.

Arnaud-Haond, Sophie, Carlos M. Duarte, Elena Diaz-Almela, Núria Marbà, Tomas Sintes, and Ester A. Serrão. "Implications of Extreme Life Span in Clonal Organisms: Millenary Clones in Meadows of the Threatened Seagrass *Posidonia oceanica.*" *PLoS One* 7, no. 2 (2012): e30454.

Boyle, Lyle. "Restoration Forth 2023 Review." The Ecology Centre, January 8, 2024. theecologycentre.org/post/restoration-forth-2023-review.

Brodersen, Kasper Elgetti, Michael Kühl, Daniel A. Nielsen, Ole Pedersen, and Anthony W. D. Larkum. "Rhizome, Root/Sediment Interactions, Aerenchyma and Internal Pressure Changes in Seagrasses." In *Seagrasses of Australia: Structure, Ecology and Conservation*, edited by Anthony W. D. Larkum, Gary A. Kendrick, and Peter J. Ralph. Springer International, 2018.

Campbell, Anthony D., Lola Fatoyinbo, Liza Goldberg, and David Lagomasino. "Global Hotspots of Salt Marsh Change and Carbon Emissions." *Nature* 612, no. 7941 (2022): 701–706.

Chen, Ling-Yun, Bei Lu, Diego F. Morales-Briones, et al. "Phylogenomic Analyses of Alismatales Shed Light into Adaptations to Aquatic Environments." *Molecular Biology and Evolution* 39, no. 5 (2022): msac079.

Chin, Diana W., Jimmy de Fouw, Tjisse van der Heide, et al. "Facilitation of a Tropical Seagrass by a Chemosymbiotic Bivalve Increases with Environmental Stress." *Journal of Ecology* 109, no. 1 (2021): 204–217.

de los Santos, Carmen B., Dorte Krause-Jensen, Teresa Alcoverro, et al. "Recent Trend Reversal for Declining European Seagrass Meadows." *Nature Communications* 10, no. 1 (2019): 3356.

Duarte, Carlos M., Iñigo J. Losada, Iris E. Hendriks, Inés Mazarrasa, and Núria Marbà. "The Role of Coastal Plant Communities for Climate Change Mitigation and Adaptation." *Nature Climate Change* 3, no. 11 (2013): 961–968.

Edgeloe, Jane M., Anita A. Severn-Ellis, Philipp E. Bayer, et al. "Extensive Polyploid Clonality Was a Successful Strategy for Seagrass to Expand into a Newly Submerged Environment." *Proceedings of the Royal Society B* 289, no. 1976 (2022): 20220538.

Fahimipour, Ashkaan K., Melissa R. Kardish, Jenna M. Lang, Jessica L. Green, Jonathan A. Eisen, and John J. Stachowicz. "Global-Scale Structure of the Eelgrass Microbiome." *Applied and Environmental Microbiology* 83, no. 12 (2017): e03391-16.

Foster, Nicole R., Eugenia T. Apostolaki, Katelyn DiBenedetto, et al. "Societal Value of Seagrass from Historical to Contemporary Perspectives." *Ambio* 54 (2025): 1–17.

Friess, Daniel A., Kerrylee Rogers, Catherine E. Lovelock, et al. "The State of the World's Mangrove Forests: Past, Present, and Future." *Annual Review of Environment and Resources* 44 (2019): 89–115.

Fu, Chuancheng, Sofia Frappi, Michelle Nicole Havlik, et al. "Substantial Blue Carbon Sequestration in the World's Largest Seagrass Meadow." *Communications Earth & Environment* 4, no. 1 (2023): 474.

Govers, Laura L., Jannes H. T. Heusinkveld, Max L. E. Gräfnings, Quirin Smeele, and Tjisse van der Heide. "Adaptive Intertidal Seed-Based Seagrass Restoration in the Dutch Wadden Sea." *PLoS One* 17, no. 2 (2022): e0262845.

Graham, Angus. "Archaeological Notes on Some Harbours in Eastern Scotland." *Proceedings of the Society of Antiquaries of Scotland* 101 (1971): 200–285.

Hensel, Marc J. S., Christopher J. Patrick, Robert J. Orth, et al. "Rise of *Ruppia* in Chesapeake Bay: Climate Change–Driven Turnover of Foundation Species Creates New Threats and Management Opportunities." *Proceedings of the National Academy of Sciences* 120, no. 23 (2023): e2220678120.

Historic Environment Scotland. "Blackness Castle Statement of Significance." Published October 1, 2016, and updated July 31, 2019. historicenvironment.scot/archives-and-research/publications/publication/?publicationId=f313e391-55fa-4f83-a8f2-a57000c31c1d.

Historic Environment Scotland. "Blackness, Harbour." canmore.org.uk/site/280044/blackness-harbour.

Hooijer, Aljosja, and Ronald Vernimmen. "Global LiDAR Land Elevation Data Reveal Greatest Sea-Level Rise Vulnerability in the Tropics." *Nature Communications* 12, no. 1 (2021): 3592.

IPCC (Core Writing Team, H. Lee, and J. Romero, eds.). *Climate Change 2023: Synthesis Report. Contribution of Working Groups I, II and III to the Sixth Assessment Report of the Intergovernmental Panel on Climate Change.* IPCC, 2023. ipcc.ch/report/ar6/syr/downloads/report/IPCC_AR6_SYR_LongerReport.pdf.

Jackson, Jeremy B. C. "Reefs Since Columbus." *Coral Reefs* 16 (1997): S23–S32.

Kendrick, Gary A., Robert J. Nowicki, Ylva S. Olsen, et al. "A Systematic Review of How

Multiple Stressors from an Extreme Event Drove Ecosystem-Wide Loss of Resilience in an Iconic Seagrass Community." *Frontiers in Marine Science* 6 (2019): 455.

Krause-Jensen, Dorte, Oscar Serrano, Eugenia T. Apostolaki, David J. Gregory, and Carlos M. Duarte. "Seagrass Sedimentary Deposits as Security Vaults and Time Capsules of the Human Past." *Ambio* 48 (2019): 325–335.

Krause-Jensen, Dorte, Carlos M. Duarte, Kaj Sand-Jensen, and Jacob Carstensen. "Century-Long Records Reveal Shifting Challenges to Seagrass Recovery." *Global Change Biology* 27, no. 3 (2021): 563–575.

Lange, Troels, Nele S. Oncken, Niels Svane, Rune C. Steinfurth, Erik Kristensen, and Mogens R. Flindt. "Large-Scale Eelgrass Transplantation: A Measure for Carbon and Nutrient Sequestration in Estuaries." *Marine Ecology Progress Series* 685 (2022): 97–109.

Larkum, Anthony W. D., Robert J. Orth, Carlos M. Duarte, eds. *Seagrasses: Biology, Ecology and Conservation.* Springer, 2006.

Leemans, Luuk, Isis Martínez, Tjisse van der Heide, Marieke M. van Katwijk, and Brigitta I. van Tussenbroek. "A Mutualism Between Unattached Coralline Algae and Seagrasses Prevents Overgrazing by Sea Turtles." *Ecosystems* 23 (2020): 1631–1642.

Macreadie, Peter I., Stacey M. Trevathan-Tackett, Charles G. Skilbeck, et al. "Losses and Recovery of Organic Carbon from a Seagrass Ecosystem Following Disturbance." *Proceedings of the Royal Society B: Biological Sciences* 282, no. 1817 (2015): 20151537.

Macreadie, Peter I., Micheli D. P. Costa, Trisha B. Atwood, et al. "Blue Carbon as a Natural Climate Solution." *Nature Reviews Earth & Environment* 2, no. 12 (2021): 826–839.

Mann, Hugo F., Natalie E. Wildermann, Chuancheng Fu, et al. "Green Turtle Tracking Leads the Discovery of Seagrass Blue Carbon Resources." *Proceedings of the Royal Society B: Biological Sciences* 291, no. 2035 (2024): 20240502.

Mazaris, Antonios D., Gail Schofield, Chrysoula Gkazinou, Vasiliki Almpanidou, and Graeme C. Hays. "Global Sea Turtle Conservation Successes." *Science Advances* 3, no. 9 (2017): e1600730.

McKenzie, Len J., Rudi L. Yoshida, John W. Aini, et al. "Seagrass Ecosystem Contributions to People's Quality of Life in the Pacific Island Countries and Territories." *Marine Pollution Bulletin* 167 (2021): 112307.

Mohr, Wiebke, Nadine Lehnen, Soeren Ahmerkamp, et al. "Terrestrial-Type Nitrogen-Fixing Symbiosis Between Seagrass and a Marine Bacterium." *Nature* 600, no. 7887 (2021): 105–109.

Olsen, Jeanine L., Pierre Rouzé, Bram Verhelst, et al. "The Genome of the Seagrass *Zostera marina* Reveals Angiosperm Adaptation to the Sea." *Nature* 530, no. 7590 (2016): 331–335.

Orth, Robert J., Tim J. B. Carruthers, William C. Dennison, et al. "A Global Crisis for Seagrass Ecosystems." *Bioscience* 56, no. 12 (2006): 987–996.

Orth, Robert J., Jonathan S. Lefcheck, Karen S. McGlathery, et al. "Restoration of Seagrass Habitat Leads to Rapid Recovery of Coastal Ecosystem Services." *Science Advances* 6, no. 41 (2020): eabc6434.

Orth, Robert J., and Kenneth L. Heck Jr. "The Dynamics of Seagrass Ecosystems: History, Past Accomplishments, and Future Prospects." *Estuaries and Coasts* 46, no. 7 (2023): 1653–1676.

Potouroglou, Maria, James C. Bull, Ken W. Krauss, et al. "Measuring the Role of Seagrasses in Regulating Sediment Surface Elevation." *Scientific Reports* 7, no. 1 (2017): 11917.

Potouroglou, Maria, Danielle Whitlock, Luna Milatovic, et al. "The Sediment Carbon stocks of Intertidal Seagrass Meadows in Scotland." *Estuarine, Coastal and Shelf Science* 258 (2021): 107442.

"Restoration Forth." World Wildlife Fund. wwf.org.uk/scotland/restoration-forth.

Rezek, Ryan J., Bradley T. Furman, Robin P. Jung, Margaret O. Hall, and Susan S. Bell. "Long-Term Performance of Seagrass Restoration Projects in Florida, USA." *Scientific Reports* 9, no. 1 (2019): 15514.

Ross, Iain. "Oysters and the Firth of Forth." Maorach Beag, Scottish Shellfish. maorach beag.scot/blogs/journal/oysters-and-the-firth-of-forth-part-1-pre-industrial -revolution.

Roth, Florian, Elias Broman, Xiaole Sun, et al. "Methane Emissions Offset Atmospheric Carbon Dioxide Uptake in Coastal Macroalgae, Mixed Vegetation and Sediment Ecosystems." *Nature Communications* 14, no. 1 (2023): 42.

Sanchez-Vidal, Anna, Miquel Canals, William P. De Haan, Javier Romero, and Marta Veny. "Seagrasses Provide a Novel Ecosystem Service by Trapping Marine Plastics." *Scientific Reports* 11, no. 1 (2021): 254.

Schorn, Sina, Soeren Ahmerkamp, Emma Bullock, et al. "Diverse Methylotrophic Methanogenic Archaea Cause High Methane Emissions from Seagrass Meadows." *Proceedings of the National Academy of Sciences* 119, no. 9 (2022): e2106628119.

Sogin, E. Maggie, Dolma Michellod, Harald R. Gruber-Vodicka, et al. "Sugars Dominate the Seagrass Rhizosphere." *Nature Ecology & Evolution* 6, no. 7 (2022): 866–877.

Speelman, Eveline N., M. M. L. Van Kempen, J. Barke, et al. "The Eocene Arctic *Azolla* Bloom: Environmental Conditions, Productivity and Carbon Drawdown." *Geobiology* 7, no. 2 (2009): 155–170.

Steinfurth, Rune C., Troels Lange, Nele S. Oncken, Erik Kristensen, Cintia O. Quintana, and Mogens R. Flindt. "Improved Benthic Fauna Community Parameters After Large-Scale Eelgrass (*Zostera marina*) Restoration in Horsens Fjord, Denmark." *Marine Ecology Progress Series* 687 (2022): 65–77.

Stevens, P. F. "Alismatales." Angiosperm Phylogeny Website. Version 14, July 2017. mobot .org/MOBOT/research/APweb.

Telesca, Luca, Andrea Belluscio, Alessandro Criscoli, et al. "Seagrass Meadows (*Posidonia oceanica*) Distribution and Trajectories of Change." *Scientific Reports* 5, no. 1 (2015): 12505.

Temmink, Ralph J. M., Leon P. M. Lamers, Christine Angelini, et al. "Recovering Wetland Biogeomorphic Feedbacks to Restore the World's Biotic Carbon Hotspots." *Science* 376, no. 6593 (2022): eabn1479.

Thomas, Bethan. *Seagrass (*Zostera marina*) in the Firth of Forth, Scotland.* Firth of Forth Seagrass Project, 2020.

Trevathan-Tackett, Stacey M., Peter I. Macreadie, Jonathan Sanderman, Jeff Baldock, Johanna M. Howes, and Peter J. Ralph. "A Global Assessment of the Chemical Recalcitrance of Seagrass Tissues: Implications for Long-Term Carbon Sequestration." *Frontiers in Plant Science* 8 (2017): 925.

United Nations Environment Programme. *Out of the Blue: The Value of Seagrasses to the Environment and to People.* UNEP, 2020.

van Tussenbroek, Brigitta I., L. Veronica Monroy-Velazquez, and Vivianne Solis-Weiss. "Meso-Fauna Foraging on Seagrass Pollen May Serve in Marine Zoophilous Pollination." *Marine Ecology Progress Series* 469 (2012): 1–6.

van Tussenbroek, Brigitta I., Nora Villamil, Judith Márquez-Guzmán, Ricardo Wong, L. Verónica Monroy-Velázquez, and Vivianne Solis-Weiss. "Experimental Evidence of Pollination in Marine Flowers by Invertebrate Fauna." *Nature Communications* 7, no. 1 (2016): 12980.

van Tussenbroek, Brigitta I., Héctor A. Hernández Arana, Rosa E. Rodríguez-Martínez, et al. "Severe Impacts of Brown Tides Caused by *Sargassum* spp. on Near-Shore Caribbean Seagrass Communities." *Marine Pollution Bulletin* 122, no. 1–2 (2017): 272–281.

Waycott, Michelle, Carlos M. Duarte, Tim J. B. Carruthers, et al. "Accelerating Loss of Seagrasses Across the Globe Threatens Coastal Ecosystems." *Proceedings of the National Academy of Sciences* 106, no. 30 (2009): 12377–12381.

Rose

Agatonovic-Kustrin, Snezana, Chloe Ke Yi Chan, Vladimir Gegechkori, and David W. Morton. "Models for Skin and Brain Penetration of Major Components from Essential Oils Used in Aromatherapy for Dementia Patients." *Journal of Biomolecular Structure and Dynamics* 38, no. 8 (2020).

Allen, Caroline, Jan Havlíček, and S. Craig Roberts. "Effect of Fragrance Use on Discrimination of Individual Body Odor." *Frontiers in Psychology* 6 (2015): 1115.

Arriaga-Osnaya, Brenda Jessica, Jorge Contreras-Garduño, Francisco Javier Espinosa-García, et al. "Are Body Size and Volatile Blends Honest Signals in Orchid Bees?" *Ecology and Evolution* 7, no. 9 (2017): 3037–3045.

L'Association de Sauvegarde de la Siagne et de son Canal. "L'histoire du canal." canaldela siagne.fr/lhistoire-du-canal.

Baldermann, Susanne, Ziyin Yang, Miwa Sakai, Peter Fleischmann, and Naoharu Watanabe. "Volatile Constituents in the Scent of Roses." *Floriculture and Ornamental Biotechnology* 3, no. 1 (2009): 89–97.

Başer, K. Hüsnü Can. "Rose Mentioned in the Works of Scientists of the Medieval East and Implications in Modern Science." *Natural Product Communications* 12, no. 8 (2017): 1327–1330.

Baudino, Sylvie, Jean-Claude Caissard, Véronique Bergougnoux, et al. "Production and Emission of Volatile Compounds by Petal Cells." *Plant Signaling & Behavior* 2, no. 6 (2007): 525–526.

Bendahmane, Mohammed, Annick Dubois, Olivier Raymond, and Manuel Le Bris. "Genetics and Genomics of Flower Initiation and Development in Roses." *Journal of Experimental Botany* 64, no. 4 (2013): 847–857.

Brand, Philipp, Ismael A. Hinojosa-Díaz, Ricardo Ayala, et al "The Evolution of Sexual Signaling Is Linked to Odorant Receptor Tuning in Perfume-Collecting Orchid Bees." *Nature Communications* 11, no. 1 (2020): 244.

Byers, Kelsey J. R. P., H. D. Bradshaw Jr., and Jeffrey A. Riffell. "Three Floral Volatiles Contribute to Differential Pollinator Attraction in Monkeyflowers (*Mimulus*)." *Journal of Experimental Biology* 217, no. 4 (2014): 614–623.

Byl, Sheila Ann. "The Essence and Use of Perfume in Ancient Egypt." PhD diss., University of South Africa, 2012.

Castel, Cécilia, Xavier Fernandez, Jean-Jacques Filippi, and Jean-Pierre Brun. "Perfumes in Mediterranean Antiquity." *Flavour and Fragrance Journal* 24, no. 6 (2009): 326–334.

"Centifolias." Historic Roses Group, August 31, 2017. historicroses.org/certifolias.

Chen, Chun, Qishi Song, Magali Proffit, Jean-Marie Bessière, Zongbo Li, and Martine Hossaert-McKey. "Private Channel: A Single Unusual Compound Assures Specific Pollinator Attraction in *Ficus semicordata*." *Functional Ecology* 23, no. 5 (2009): 941–950.

Classen, Constance, David Howes, and Anthony Synnott. *Aroma: The Cultural History of Smell*. Routledge, 1994.

Cobb, Matthew. *Smell: A Very Short Introduction*. Oxford University Press, 2020.

Conart, Corentin, Dikki Pedenla Bomzan, Xing-Qi Huang, et al. "A Cytosolic Bifunctional Geranyl/Farnesyl Diphosphate Synthase Provides MVA-Derived GPP for Geraniol Biosynthesis in Rose Flowers." *Proceedings of the National Academy of Sciences* 120, no. 19 (2023): e2221440120.

Cousin, Laura. "Beauty Experts: Female Perfume-Makers in the 1st Millennium BC." In *The Role of Women in Work and Society in the Ancient Near East*, edited by Brigitte Lion and Cécile Michel. De Gruyter, 2016.

de Lacy Costello, Ben, Anton Amann, Huda Al-Kateb, et al. "A Review of the Volatiles from the Healthy Human Body." *Journal of Breath Research* 8, no. 1 (2014): 014001.

De Sousa, Hilário. "Changes in the Language of Perception in Cantonese." *Senses and Society* 6, no. 1 (2011): 38–47.

Dobson, Heidi E. M., Juan Arroyo, Gunnar Bergström, and Inga Groth. "Interspecific Variation in Floral Fragrances Within the Genus *Narcissus* (Amaryllidaceae)." *Biochemical Systematics and Ecology* 25, no. 8 (1997): 685–706.

Dobson, Heidi E. M. "Relationship Between Floral Fragrance Composition and Type of Pollinator." In *Biology of Floral Scent*, edited by Natalia Dudareva and Eran Pichersky. CRC Press, 2006.

Dötterl, Stefan, and Jonathan Gershenzon. "Chemistry, Biosynthesis and Biology of Floral Volatiles: Roles in Pollination and Other Functions." *Natural Product Reports* 40, no. 12 (2023): 1901–1937.

Dudareva, Natalia, and Eran Pichersky, eds. *Biology of Floral Scent.* Taylor and Francis, 2006.

Eltz, Thomas, Andreas Sager, and Klaus Lunau. "Juggling with Volatiles: Exposure of Perfumes by Displaying Male Orchid Bees." *Journal of Comparative Physiology A* 191 (2005): 575–581.

Eltz, Thomas, Yvonne Zimmermann, Carolin Pfeiffer, et al. "An Olfactory Shift Is Associated with Male Perfume Differentiation and Species Divergence in Orchid Bees." *Current Biology* 18, no. 23 (2008): 1844–1848.

Farré-Armengol, Gerard, Josep Peñuelas, Tao Li, et al. "Ozone Degrades Floral Scent and Reduces Pollinator Attraction to Flowers." *New Phytologist* 209, no. 1 (2016): 152–160.

Farré-Armengol, Gerard, Marcos Fernández-Martínez, Iolanda Filella, Robert R. Junker, and Josep Peñuelas. "Deciphering the Biotic and Climatic Factors That Influence Floral Scents: A Systematic Review of Floral Volatile Emissions." *Frontiers in Plant Science* 11 (2020): 555907.

Ferdenzi, Camille, Stéphane Richard Ortegón, Sylvain Delplanque, Nicolas Baldovini, and Moustafa Bensafi. "Interdisciplinary Challenges for Elucidating Human Olfactory Attractiveness." *Philosophical Transactions of the Royal Society B* 375, no. 1800 (2020): 20190268.

Folkard, Richard. *Plant Lore, Legends, and Lyrics.* London: Sampson Low, Marston, Searle, and Rivington, 1884. gutenberg.org/ebooks/44638.

Forestell, Catherine A. "Does Maternal Diet Influence Future Infant Taste and Odor Preferences? A Critical Analysis." *Annual Review of Nutrition* 44 (2024).

Friberg, Magne, Christopher Schwind, Paulo R. Guimarães Jr., Robert A. Raguso, and John N. Thompson. "Extreme Diversification of Floral Volatiles Within and Among Species of *Lithophragma* (Saxifragaceae)." *Proceedings of the National Academy of Sciences* 116, no. 10 (2019): 4406–4415.

Gerard, John. *Herball, or Generall Historie of Plantes.* London: Bonham and John Norton, 1597. biodiversitylibrary.org/bibliography/51606.

Gerkin, Richard C., and Jason B. Castro. "The Number of Olfactory Stimuli That Humans Can Discriminate Is Still Unknown." *eLife* 4 (2015): e08127.

Godin, Mélissa. "Climate Crisis Brings Whiff of Danger to French Perfume Capital." *The Guardian*, February 18, 2023. theguardian.com/world/2023/feb/18/climate-crisis-brings-whiff-of-danger-to-french-perfume-capital.

Gonzalez-Terrazas, Tania P., Carlos Martel, Paulo Milet-Pinheiro, Manfred Ayasse, Elisabeth K. V. Kalko, and Marco Tschapka. "Finding Flowers in the Dark: Nectar-Feeding Bats Integrate Olfaction and Echolocation While Foraging for Nectar." *Royal Society Open Science* 3, no. 8 (2016): 160199.

Hämmerli, August, C. Schweisgut, and Manuel Kaegi. "Population Genetic Segmentation of MHC-Correlated Perfume Preferences." *International Journal of Cosmetic Science* 34, no. 2 (2012): 161–168.

Hardy, Karen. "Paleomedicine and the Evolutionary Context of Medicinal Plant Use." *Revista Brasileira de Farmacognosia* 31 (2021): 1–15.

Havlíček, Jan, and S. Craig Roberts. "The Perfume-Body Odour Complex: An Insightful Model for Culture–Gene Coevolution?" In *Chemical Signals in Vertebrates 12*, edited by Marion L. East and Martin Dehnhard. Springer, 2012.

Heil, Martin. "Indirect Defence via Tritrophic Interactions." *New Phytologist* 178, no. 1 (2008): 41–61.

Henske, Jonas, Nicholas W. Saleh, Thomas Chouvenc, Santiago R. Ramírez, and Thomas Eltz. "Function of Environment-Derived Male Perfumes in Orchid Bees." *Current Biology* 33, no. 10 (2023): 2075–2080.

Henske, Jonas, and Thomas Eltz. "Age-Dependent Perfume Development in Male Orchid Bees, *Euglossa imperialis*." *Journal of Experimental Biology* 227, no. 6 (2024): jeb246995.

Heymann, Eckhard W. "Florivory, Nectarivory, and Pollination—a Review of Primate-Flower Interactions." *Ecotropica* 17 (2011): 41–52.

Hogan, Jeremy D., Amanda D. Melin, Krisztina N. Mosdossy, and Linda M. Fedigan. "Seasonal Importance of Flowers to Costa Rican Capuchins (*Cebus capucinus imitator*): Implications for Plant and Primate." *American Journal of Physical Anthropology* 161, no. 4 (2016): 591–602.

Hůla, Martin, and Jaroslav Flegr. "Habitat Selection and Human Aesthetic Responses to Flowers." *Evolutionary Human Sciences* 3 (2021): e5.

Inogwabini, Bila-Isia, and Bewa Matungila. "Bonobo Food Items, Food Availability and Bonobo Distribution in the Lake Tumba Swampy Forests, Democratic Republic of Congo." *Open Conservation Biology Journal* 3, no. 1 (2009): 14–23.

Iwata, Hikaru, Tsuneo Kato, and Susumu Ohno. "Triparental Origin of Damask Roses." *Gene* 259, no. 1–2 (2000): 53–59.

Jardins du Musée International de la Parfumerie. Booklet published by Communauté d'Agglomération du Pays de Grasse, 2018.

Jena, Jitendra, Tripathi Vineeta, Ashok Kumar, Kumar Brijesh, and Pankaj Singh. "Rosa Centifolia: Plant Review." *International Journal of Research in Pharmacy and Chemistry* 2, no. 3 (2012): 794–796.

Joo, Seong Soo, Yun-Bae Kim, and Do Ik Lee. "Antimicrobial and Antioxidant Properties of Secondary Metabolites from White Rose Flower." *Plant Pathology Journal* 26, no. 1 (2010): 57–62.

Junker, Robert R., and Nico Blüthgen. "Floral Scents Repel Potentially Nectar-Thieving Ants." *Evolutionary Ecology Research* 10, no. 2 (2008): 295–308.

Kessler, Danny, Celia Diezel, and Ian T. Baldwin. "Changing Pollinators as a Means of Escaping Herbivores." *Current Biology* 20, no. 3 (2010): 237–242.

Kimsey, Lynn Siri. "The Behaviour of Male Orchid Bees (Apidae, Hymenoptera, Insecta) and the Question of Leks." *Animal Behaviour* 28, no. 4 (1980): 996–1004.

King, Anya. "Medieval Islamicate Aromatherapy: Medical Perspectives on Aromatics and Perfumes." *Senses and Society* 17, no. 1 (2022): 37–51.

Kockmann, Norbert. "History of Distillation." In *Distillation*, edited by Andrzej Górak and Eva Sorensen. Academic Press, 2014.

Lenochová, Pavlína, Pavla Vohnoutova, S. Craig Roberts, Elisabeth Oberzaucher, Karl Grammer, and Jan Havlíček. "Psychology of Fragrance Use: Perception of Individual Odor and Perfume Blends Reveals a Mechanism for Idiosyncratic Effects on Fragrance Choice." *PLoS One* 7, no. 3 (2012): e33810.

Liu, Jasen W., Paulo Milet-Pinheiro, Günter Gerlach, et al. "Macroevolution of Floral Scent Chemistry Across Radiations of Male Euglossine Bee-Pollinated Plants." *Evolution* 78, no. 1 (2024): 98–110.

Majid, Asifa, and Nicole Kruspe. "Hunter-Gatherer Olfaction Is Special." *Current Biology* 28, no. 3 (2018): 409–413.

McGee, Harold. *Nose Dive: A Field Guide to the World's Smells.* Penguin Press, 2020.

McGrew, William C., P. J. Baldwin, and C. E. G. Tutin. "Diet of Wild Chimpanzees (*Pan troglodytes verus*) at Mt. Assirik, Senegal: I. Composition." *American Journal of Primatology* 16, no. 3 (1988): 213–226.

McPhee, Peter. *Living the French Revolution, 1789–99.* Palgrave Macmillan, 2006.

Meister, Markus. "On the Dimensionality of Odor Space." *eLife* 4 (2015): e07865.

Mostafa, Salma, Yun Wang, Wen Zeng, and Biao Jin. "Floral Scents and Fruit Aromas: Functions, Compositions, Biosynthesis, and Regulation." *Frontiers in Plant Science* 13 (2022): 860157.

Moutinho, Sofia. "Why Cats Are Crazy for Catnip." *Science,* January 20, 2021. science.org /content/article/why-cats-are-crazy-catnip.

Nadel, Dani, Avinoam Danin, Robert C. Power, et al. "Earliest Floral Grave Lining from 13,700–11,700-Y-Old Natufian Burials at Raqefet Cave, Mt. Carmel, Israel." *Proceedings of the National Academy of Sciences* 110, no. 29 (2013): 11774–11778.

Nasir, Mubashir Hussain, Raziya Nadeem, Kalsoom Akhtar, Muhammad Asif Hanif, and Ahmad M. Khalid. "Efficacy of Modified Distillation Sludge of Rose (*Rosa centifolia*) Petals for Lead (II) and Zinc (II) Removal from Aqueous Solutions." *Journal of Hazardous Materials* 147, no. 3 (2007): 1006–1014.

Pearce, Richard F., Luca Giuggioli, and Sean A. Rands. "Bumblebees Can Discriminate Between Scent-Marks Deposited by Conspecifics." *Scientific Reports* 7, no. 1 (2017): 43872.

Peters, Ruud, Rick Veenstra, Karin Heutinck, Albert Baas, Sandra Munniks, and Jaap Knotter. "Human Scent Characterization: A Review." *Forensic Science International* 349 (2023): 111743.

Pichersky, Eran, Joseph P. Noel, and Natalia Dudareva. "Biosynthesis of Plant Volatiles: Nature's Diversity and Ingenuity." *Science* 311, no. 5762 (2006): 808–811.

Pliny the Elder. "Book XIII. The Natural History of Exotic Trees, and an Account of Unguents." In *The Natural History of Pliny*, vol. 3. Translated by John Bostock and Henry T. Riley. London: Henry G. Bond, 1855. gutenberg.org/files/59131/59131-h/59131-h.htm #Page_168.

Rageot, Maxime, Ramadan B. Hussein, Susanne Beck, et al. "Biomolecular Analyses Enable New Insights into Ancient Egyptian Embalming." *Nature* 614, no. 7947 (2023): 287–293.

Ramírez, Santiago R., Thomas Eltz, Mikiko K. Fujiwara, et al. "Asynchronous Diversification in a Specialized Plant-Pollinator Mutualism." *Science* 333, no. 6050 (2011): 1742–1746.

Ramirez, Wendy, Carolina Gomez, Nyasha K. T. Thomas, Imar J. Muktar, and Olena Riabinina. "The Neuroecology of Olfaction in Bees." *Current Opinion in Insect Science* 56 (2023): 101018.

Ramos, Sergio E., and Florian P. Schiestl. "Evolution of Floral Fragrance Is Compromised by Herbivory." *Frontiers in Ecology and Evolution* 8 (2020): 30.

Rasmann, Sergio, Tobias G. Köllner, Jörg Degenhardt, et al. "Recruitment of Entomopathogenic Nematodes by Insect-Damaged Maize Roots." *Nature* 434, no. 7034 (2005): 732–737.

Reinhard, Judith, and Mandyam V. Srinivasan. "The Role of Scents in Honey Bee Foraging and Recruitment." In *Food Exploitation by Social Insects: Ecological, Behavioral, and Theoretical Approaches*, edited by Stefan Jarau and Michael Hrncir. CRC Press, 2009.

Riba-Hernández, Pablo, and Kathryn E. Stoner. "Massive Destruction of *Symphonia globulifera* (Clusiaceae) Flowers by Central American Spider Monkeys (*Ateles geoffroyi*)." *Biotropica* 37, no. 2 (2005): 274–278.

"Roses." Observatory of Economic Complexity, 2024. oec.world/en/profile/hs/roses.

Ross, Abigail C., Margaret A. H. Bryer, Colin A. Chapman, Jessica M. Rothman, Omer Nevo, and Kim Valenta. "Why Eat Flowers? *Symphonia globulifera* Flowers Provide a Fatty Resource for Red-Tailed Monkeys." *Folia Primatologica* 93, no. 1 (2022): 41–52.

Roubik, David W., and Jette T. Knudsen. "An Embellishment That Became a Mutualism: Inquiries on Male Bee Tibial Bouquets and Fragrance-Producing Orchids in Panama and Oceanic Islands (Apidae: Apinae, Euglossini; Orchidaceae: Epidendroideae)." *Flora* 232 (2017): 117–127.

Scalliet, Gabriel, Florence Piola, Christophe J. Douady, et al. "Scent Evolution in Chinese Roses." *Proceedings of the National Academy of Sciences* 105, no. 15 (2008): 5927–5932.

Shabbir, M. Khalid, Raziya Nadeem, Hamid Mukhtar, Farooq Anwar, and Muhammad W. Mumtaz. "Physico-Chemical Analysis and Determination of Various Chemical Constituents of Essential Oil in *Rosa centifolia*." *Pakistan Journal of Botany* 41, no. 2 (2009): 615–620.

Shalit, Moshe, Inna Guterman, Hanne Volpin, et al. "Volatile Ester Formation in Roses. Identification of an Acetyl-Coenzyme A. Geraniol/Citronellol Acetyltransferase in Developing Rose Petals." *Plant Physiology* 131, no. 4 (2003): 1868–1876.

Sorokowska, Agnieszka, Piotr Sorokowski, and Jan Havlíček. "Body Odor Based Personality Judgments: The Effect of Fragranced Cosmetics." *Frontiers in Psychology* 7 (2016): 530.

Sowndhararajan, Kandhasamy, and Songmun Kim. "Influence of Fragrances on Human Psychophysiological Activity: With Special Reference to Human Electroencephalographic Response." *Scientia Pharmaceutica* 84, no. 4 (2016): 724–752.

Spahn, Joanne M., Emily H. Callahan, Maureen K. Spill, et al. "Influence of Maternal Diet on Flavor Transfer to Amniotic Fluid and Breast Milk and Children's Responses: A Systematic Review." *American Journal of Clinical Nutrition* 109 (2019): 1003S–1026S.

Stevens, Wallace. "Bouquet of Roses in Sunlight." *Poetry* 74, no. 1 (1947). poetryfoundation.org/poetrymagazine/browse?contentId=24852.

Thompson, R. Campbell. "Assyrian Prescriptions for Diseases of the Chest and Lungs." *Revue d'Assyriologie et d'archéologie orientale* 31, no. 1 (1934): 1–29.

Trimmer, Casey, Andreas Keller, Nicolle R. Murphy, et al. "Genetic Variation Across the Human Olfactory Receptor Repertoire Alters Odor Perception." *Proceedings of the National Academy of Sciences* 116, no. 19 (2019): 9475–9480.

UNESCO Intangible Cultural Heritage. "The Skills Related to Perfume in Pays de Grasse." 2018. ich.unesco.org/en/RL/the-skills-related-to-perfume-in-pays-de-grasse

-the-cultivation-of-perfume-plants-the-knowledge-and-processing-of-natural-raw -materials-and-the-art-of-perfume-composition-01207.

Ville de Grasse. "Grasse, la source de La Foux." Fiche Patrimoine no. 23. ville-grasse.fr /wp-content/uploads/2024/10/FICHE-PATRIMOINE-24-LA-FOUX.pdf.

Wadley, Lyn, Christine Sievers, Marion Bamford, Paul Goldberg, Francesco Berna, and Christopher Miller. "Middle Stone Age Bedding Construction and Settlement Patterns at Sibudu, South Africa." *Science* 334, no. 6061 (2011): 1388–1391.

Watts, David P., Kevin B. Potts, Jeremiah S. Lwanga, and John C. Mitani. "Diet of Chimpanzees (Pan troglodytes schweinfurthii) at Ngogo, Kibale National Park, Uganda, 1. Diet Composition and Diversity." *American Journal of Primatology* 74, no. 2 (2012): 114–129.

Weber, M. G., L. Mitko, T. Eltz, and S. R. Ramírez, 2016. Macroevolution of Perfume Signalling in Orchid Bees. *Ecology Letters* 19, no. 11 (2016): 1314–1323.

Wright, Geraldine A., and Florian P. Schiestl. "The Evolution of Floral Scent: The Influence of Olfactory Learning by Insect Pollinators on the Honest Signalling of Floral Rewards." *Functional Ecology* 23, no. 5 (2009): 841–851.

Tea

Berwick, Leonie, and Isabelle Charmantier, eds. *L: 50 Objects, Stories & Discoveries from the Linnean Society of London*. Linnean Society of London, 2020.

Blunt, Wilfrid. *Linnaeus: The Compleat Naturalist*. Princeton University Press, 2001.

Bremer, Birgitta. "Linnaeus' Sexual System and Flowering Plant Phylogeny." *Nordic Journal of Botany* 25, no. 1–2 (2007): 5–6.

Charmantier, Isabelle. "Linnaeus and Race." Linnean Society of London, September 3, 2020. linnean.org/learning/who-was-linnaeus/linnaeus-and-race.

Darwin, C. R. *The Descent of Man, and Selection in Relation to Sex*. Vol. 1. London: John Murray, 1871. darwin-online.org.uk/EditorialIntroductions/Freeman_TheDescentofMan.html.

Darwin, Francis, ed. *The Life and Letters of Charles Darwin, Including an Autobiographical Chapter*. Vol. 3. London: John Murray, 1887. darwin-online.org.uk.

de Jussieu, Antoine. "Histoire du Café." Histoire de l'Académie royale des sciences. Paris: Imprimerie Royale, 1715 (1719 printed edition).

Deligorges, Stéphane, Alexandre Gady, and Françoise Labalette. *Le Jardin des plantes et le Muséum national d'histoire naturalle*. Éditions du Patrimoine—Centre des Monuments Nationaux, 2004.

Gillman, Len Norman, and Shane Donald Wright. "Restoring Indigenous Names in Taxonomy." *Communications Biology* 3, no. 1 (2020): 609.

Koerner, Lisbet. *Linnaeus: Nature and Nation*. Harvard University Press, 1999.

Landweber, Julia. "'This Marvelous Bean': Adopting Coffee into Old Regime French Culture and Diet." *French Historical Studies* 38, no. 2 (2015): 193–223.

Linnaeus, Carl. *Systema Naturae*. Leiden, the Netherlands, 1735. doi.org/10.5962/bhl .title.877.

Linnaeus, Carl. *Musa Cliffortiana. Clifford's Banana Plant*. Reprint and translation of the original edition (Leiden, 1736). Translated by Stephen Freer. A. R. G. Ganter Verlag, 2007.

Linnaeus, Carl. *Hortus Cliffortianus*. Amsterdam, 1737. Linnaean Society of London, ref. no. BL.1186. linnean.access.preservica.com/uncategorized/IO_fd050fd9-4223-46d2 -9e9c-91cac4076f52.

Linnaeus, Carl. *Systema Naturae*. 10th ed. 1758. Linnaean Society of London, ref. No. BL.16/1. linnean.access.preservica.com/uncategorized/IO_936cea22-ec51-4768-a7df -452bbea1c062.

Marks, Jonathan. "Long Shadow of Linnaeus's Human Taxonomy." *Nature* 447, no. 7140 (2007): 28.

Müller-Wille, Staffan, and Karen Reeds. "A Translation of Carl Linnaeus's Introduction to *Genera plantarum* (1737)." *Studies in History and Philosophy of Science Part C: Studies in History and Philosophy of Biological and Biomedical Sciences* 38, no. 3 (2007): 563–572.

Porter, Catherine, Constant Méheut, Matt Apuzzo, and Selam Gebrekidan. "The Ransom. The Root of Haiti's Misery: Reparations to Enslavers." *New York Times*, May 20, 2022. www.nytimes.com/2022/05/20/world/americas/haiti-history-colonized-france .html.

Roberts, Jason. *Every Living Thing: The Great and Deadly Race to Know All Life.* Random House, 2024.

Robin, Nicolas. "The Influence of Scientific Theories on the Design of Botanical Gardens Around 1800." *Studies in the History of Gardens & Designed Landscapes* 28, no. 3–4 (2008): 382–399.

Smellie, W., ed. *Encyclopædia Britannica: or, A Dictionary of Arts and Sciences, Compiled by a Society of Gentlemen in Scotland.* Vol. 1. Edinburgh: Encyclopaedia Britannica, 1773.

Stevens, Peter F. *The Development of Biological Systematics: Antoine-Laurent de Jussieu, Nature, and the Natural System.* Columbia University Press, 1994.

Pansy

A Catalogue of Flowers, Plants, Trees, &c. Sold by Goring & Wright, Successors to Maddock & Son, Florists, at Walworth, near London. London, 1798. Royal Horticultural Society digital archive. collections.rhs.org.uk/view/40120/1798-a-catalogue-of-flowers-plants-trees -c-sold-by-goring-wright-successors-to-maddock-son-florists-at-walworth-near -london.

"Alan Titchmarsh MBE VMH DL—Written Evidence (HSI0093)." UK Parliament, June 26, 2023. committees.parliament.uk/writtenevidence/122315/pdf.

Annual Report and Consolidated Financial Statements for the Year Ended 31 January 2024. Royal Horticultural Society, 2024. rhs.org.uk/about-us/pdfs/about-the-rhs/mission-and -strategy/past-annual-reports/rhs-annual-report-2023-2024.pdf.

Ashbee, Natalie. "Shows Sustainability." Royal Horticultural Society. rhs.org.uk/about -us/what-we-do/policies/shows-sustainability.

Baldock, Katherine C. R., Mark A. Goddard, Damien M. Hicks, et al. "Where Is the UK's Pollinator Biodiversity? The Importance of Urban Areas for Flower-Visiting Insects." *Proceedings of the Royal Society B: Biological Sciences* 282, no. 1803 (2015): 20142849.

Bentley, Sarah. "Chemical Companies 'Misleading' Gardeners over Toxic Pesticides." *The Ecologist*, June 29, 2011. theecologist.org/2011/jun/29/chemical-companies-mis leading-gardeners-over-toxic-pesticides.

Botías, Cristina, Arthur David, Elizabeth M. Hill, and Dave Goulson. "Contamination of Wild Plants Near Neonicotinoid Seed-Treated Crops, and Implications for Non-Target Insects." *Science of the Total Environment* 566–567 (2016): 269–278.

Brooklyn Bridge Park. "Horticulture." brooklynbridgepark.org/plants-wildlife/horti culture.

Brooklyn Bridge Park. "Waterfront History." brooklynbridgepark.org/about/history.

Catalogue of Plants and Seeds, Sold by Kennedy and Lee, Nursery and Seedsmen, at the Vineyard, Hammersmith. London, 1774. Wageningen University and Research eDepot. edepot .wur.nl/473307.

"Charles Bennet 3rd Earl of Tankerville." Centre for the Study of the Legacies of British Slavery. wwwdepts-live.ucl.ac.uk/lbs/person/view/2146666791.

Chwoyka, Cecily, Dominik Linhard, Thomas Durstberger, and Johann G. Zaller. "Ornamental Plants as Vectors of Pesticide Exposure and Potential Threat to Biodiversity and Human Health." *Environmental Science and Pollution Research* 31, no. 36 (2024): 49079–49099.

Classen, Constance. *Worlds of Sense: Exploring the Senses in History and Across Cultures.* Routledge, 1993.

Comba, Livio, Sarah A. Corbet, A. Barron, et al. "Garden Flowers: Insect Visits and the Floral Reward of Horticulturally-Modified Variants." *Annals of Botany* 83, no. 1 (1999): 73–86.

Conrad, Kelvin F., Martin S. Warren, Richard Fox, Mark S. Parsons, and Ian P. Woiwod. "Rapid Declines of Common, Widespread British Moths Provide Evidence of an Insect Biodiversity Crisis." *Biological Conservation* 132, no. 3 (2006): 279–291.

Corbet, Sarah A., Jennie Bee, Kanchon Dasmahapatra, et al. "Native or Exotic? Double or Single? Evaluating Plants for Pollinator-Friendly Gardens." *Annals of Botany* 87, no. 2 (2001): 219–232.

Culley, Theresa M., and Tziporah H. Feldman. "The Role of Horticulture in Plant Invasions in the Midwestern United States." *International Journal of Plant Sciences* 184, no. 4 (2023): 260–270.

The Cultural Landscape Foundation. "What We Can Do Part 2—Biological Diversity Is Important as Social Diversity with Rebecca McMackin." December 14, 2021. youtube .com/watch?v=91_EXdE8tJQ.

Cumo, Christopher. *Encyclopedia of Cultivated Plants: From Acacia to Zinnia.* Bloomsbury, 2013.

Darwin, Charles. *The Variation of Animals and Plants Under Domestication.* Vol. 1. New York: Orange Judd, 1868. darwin-online.org.uk/converted/published/1868_Variation_F879 .1/1868_Variation_F879.1.html.

Don, Monty. "Creative Cottage Style." *Gardener's World,* June 2023.

Don, Monty. "The Full Monty." *Gardener's World,* July 2023.

Du, Xiaohua, Mengye Wang, Aneta Słomka, and Huichao Liu. "Karyologic and Heterosis Studies of the Artificial Inter- and Intraspecific Hybrids of *Viola ×wittrockiana* and *Viola cornuta.*" *HortScience* 53, no. 9 (2018): 1300–1305.

Edwards, Collin B., Elise F. Zipkin, Erica H. Henry, et al. "Rapid Butterfly Declines Across the United States During the 21st Century." *Science* 387, no. 6738 (2025): 1090–1094.

Erickson, E., H. M. Patch, and C. M. Grozinger. "Herbaceous Perennial Ornamental Plants Can Support Complex Pollinator Communities." *Scientific Reports* 11, no. 1 (2021): 17352.

Forister, Matthew L., Emma M. Pelton, and Scott H. Black. "Declines in Insect Abundance and Diversity: We Know Enough to Act Now." *Conservation Science and Practice* 1, no. 8 (2019): e80.

Fox, Richard, T. M. Brereton, J. Asher, et al. *The State of the UK's Butterflies 2015.* Butterfly Conservation and the Centre for Ecology & Hydrology, 2015.

Friedman, Elizabeth, Marnie F. Hazlehurst, Christine Loftus, Catherine Karr, Kelsey N. McDonald, and Jose Ricardo Suarez-Lopez. "Residential Proximity to Greenhouse Agriculture and Neurobehavioral Performance in Ecuadorian Children." *International Journal of Hygiene and Environmental Health* 223, no. 1 (2020): 220–227.

Garbuzov, Mihail, and Francis L. W. Ratnieks. "Quantifying Variation Among Garden Plants in Attractiveness to Bees and Other Flower-Visiting Insects." *Functional Ecology* 28, no. 2 (2014): 364–374.

"Gardening Statistics 2024." RubyHome, September 5, 2022. rubyhome.com/blog/gardening-stats.

Garden Masterclass. "Rebecca McMackin in Conversation." December 8, 2023. youtube
.com/watch?v=7iSSOw5PuKQ.

Ghosh, Amitav. *Smoke and Ashes. Opium's Hidden Histories.* Farrar, Straus and Giroux, 2023.

Ginn, Franklin. "Dig for Victory! New Histories of Wartime Gardening in Britain." *Journal of Historical Geography* 38, no. 3 (2012): 294–305.

Goulson, Dave, and 232 signatories. "Call to Restrict Neonicotinoids." *Science* 360, no. 6392 (2018): 973.

Hall, D. M., G. R. Camilo, R. K. Tonietto, et al. "The City as a Refuge for Insect Pollinators." *Conservation Biology* 31, no. 1 (2017): 24–29.

Halsch, Christopher A., Sarah M. Hoyle, Aimee Code, James A. Fordyce, and Matthew L. Forister. "Milkweed Plants Bought at Nurseries May Expose Monarch Caterpillars to Harmful Pesticide Residues." *Biological Conservation* 273 (2022): 109699.

Hopwood, Jennifer, Aimee Code, Mace Vaughan, et al. *How Neonicotinoids Can Kill Bees.* Xerces Society for Invertebrate Conservations, 2016.

Jones, Jodie. "Does the Chelsea Flower Show Have a Waste Problem?" *Gardens Illustrated,* May 4, 2024.

Khachatryan, Hayk, Xuan Wei, and Alicia Rihn. "Pest Management Practices in the US Ornamental Horticulture Industry: Use of Neonicotinoid and Non-Neonicotinoid Insecticides: FE1101/FE1101, 7/2021." *EDIS* 2021, no. 4 (2021).

Krischik, Vera, Mary Rogers, Garima Gupta, and Aruna Varshney. "Soil-Applied Imidacloprid Translocates to Ornamental Flowers and Reduces Survival of Adult *Coleomegilla maculata, Harmonia axyridis*, and *Hippodamia convergens* Lady Beetles, and Larval *Danaus plexippus* and *Vanessa cardui* Butterflies." *PLoS One* 10, no. 3 (2015): e0119133.

Lentola, Andrea, A. David, A. Abdul-Sada, Andrea Tapparo, Dave Goulson, and E. M. Hill. "Ornamental Plants on Sale to the Public Are a Significant Source of Pesticide Residues with Implications for the Health of Pollinating Insects." *Environmental Pollution* 228 (2017): 297–304.

Litt, Andrea R., Adam B. Mitchell, and Douglas W. Tallamy. "Alien Plants and Insect Diversity." In *Biological Invasions and Global Insect Decline*, edited by Jonatan Rodríguez, Petr Pyšek, and Ana Novoa. Academic Press, 2024.

Marcussen, Thomas, Harvey E. Ballard, Jiří Danihelka, Ana R. Flores, Marcela V. Nicola, and John M. Watson. "A Revised Phylogenetic Classification for *Viola* (Violaceae)." *Plants* 11, no. 17 (2022): 2224.

McDonald, Miller B., and Francis Y. Kwong, eds. *Flower Seeds: Biology and Technology.* CABI, 2005.

Milesi, Cristina, Steven W. Running, Christopher D. Elvidge, John B. Dietz, Benjamin T. Tuttle, and Ramakrishna R. Nemani. "Mapping and Modeling the Biogeochemical Cycling of Turf Grasses in the United States." *Environmental Management* 36 (2005): 426–438.

"Nouvelles Ephemerides, Économiques, Seconde Partie, Analyses, et Critiques Raisonnées. N° Premier. Éloge Historique De M. Quesnay, Contenant L'Analyse de Ses Ouvrages, Par M. le Cte d'A***." www.taieb.net/auteurs/Quesnay/albon.html.

Observatory of Economic Complexity. "Cut Flowers." 2024. oec.world/en/profile/hs/cut-flowers.

Observatory of Economic Complexity. "Live Trees, Plants, Bulbs, Cut Flowers, & Ornamental Foliage." 2024. oec.world/en/profile/hs/live-trees-plants-bulbs-cut-flowers-ornamental-foliage.

Pereira, Patrícia C. G., Cláudio E. T. Parente, Gabriel O. Carvalho, et al. "A Review on Pesticides in Flower Production: A Push to Reduce Human Exposure and Environmental Contamination." *Environmental Pollution* 289 (2021): 117817.

Porseryd, Tove, Kristina Volkova Hellström, and Patrik Dinnétz. "Pesticide Residues in Ornamental Plants Marketed as Bee Friendly: Levels in Flowers, Leaves, Roots and Soil." *Environmental Pollution* 345 (2024): 123466.

Richard, Josepha Charlotte. "The Hong Merchant's Gardens During the Canton System and the Aftermath of the Opium Wars." PhD thesis, University of Sheffield, UK, 2017.

Richard, Melissa, Douglas W. Tallamy, and Adam B. Mitchell. "Introduced Plants Reduce Species Interactions." *Biological Invasions* 21 (2019): 983–992.

Royal Horticultural Society. "The RHS Sustainability Strategy." rhs.org.uk/science/sustainability/sustainability-strategy-document.

Salisbury, Andrew, James Armitage, Helen Bostock, Joe Perry, Mark Tatchell, and Ken Thompson. "Enhancing Gardens as Habitats for Flower-Visiting Aerial Insects (Pollinators): Should We Plant Native or Exotic Species?" *Journal of Applied Ecology* 52, no. 5 (2015): 1156–1164.

Salisbury, Andrew, Sarah Al-Beidh, James Armitage, et al. "Enhancing Gardens as Habitats for Plant-Associated Invertebrates: Should We Plant Native or Exotic Species?" *Biodiversity and Conservation* 26, no. 11 (2017): 2657–2673.

San Fratello, David, Benjamin L. Campbell, William G. Secor, and Julie H. Campbell. "Impact of the COVID-19 Pandemic on Gardening in the United States: Postpandemic Expectations." *HortTechnology* 32, no. 1 (2022): 32–38.

Science Media Centre. "Expert Reaction to EU Ban on Outdoor Use of Three Neonicotinoid Pesticides." April 27, 2018. sciencemediacentre.org/expert-reaction-to-eu-ban-on-outdoor-use-of-three-neonicotinoid-pesticides.

Select Vegetable and Fower Seeds, Seed Potatos, Garden Tools & All Garden Requisites. Chester, UK: Dicksons Nurseries, Seed Growers and Nurserymen, 1899. Royal Horticultural Society digital archive. collections.rhs.org.uk/view/201553/1899-select-vegetable-and-fower-seeds-seed-potatos-garden-tools-all-garden-requisites-by-dicksons-nurseries-seed-growers-and-nurserymen-chester.

Sinclair, J., and J. Freeman. *A History and Description of the Different Varieties of the Pansey, or Heartsease, Now in Cultivation in the British Gardens; Illustrated with Twenty-Four Coloured Figures, of the Choicest Sorts.* London: Effingham Wilson, 1835.

Siviter, Harry, Gabriella L. Pardee, Nicolas Baert, Scott McArt, Shalene Jha, and Felicity Muth. "Wild Bees Are Exposed to Low Levels of Pesticides in Urban Grasslands and Community Gardens." *Science of the Total Environment* 858 (2023): 159839.

Smith, Chloë. 2010. *London: Garden City?* London Wildlife Trust, Greenspace Information for Greater London, and the Greater London Authority, 2010. lbp.org.uk/downloads/Publications/HabitatInfo/LondonGardenCity.pdf.

Solnit, Rebecca. *Orwell's Roses.* Viking, 2021.

Stokstad, Erik. "European Union Expands Ban of Three Neonicotinoid Pesticides." *Science*, April 27, 2018. science.org/content/article/european-union-expands-ban-three-neonicotinoid-pesticides.

Titchmarsh, Alan "The Chelsea Flower Show Needs to Stop Pandering to Trends and Remember That It's a Celebration of Gardening." *Country Life*, May 21, 2023. countrylife.co.uk/gardens/gardening-tips/alan-titchmarsh-the-chelsea-flower-show-needs-to-stop-pandering-to-trends-and-remember-that-its-a-celebration-of-gardening-255918.

Toumi, Khaoula, Christiane Vleminckx, Joris Van Loco, and Bruno Schiffers. "Pesticide Residues on Three Cut Flower Species and Potential Exposure of Florists in Belgium." *International Journal of Environmental Research and Public Health* 13, no. 10 (2016): 943.

United Kingdom Office for National Statistics. "Census 2021: One in Eight British Households Has No Garden." May 14, 2020. ons.gov.uk/economy/environmentalaccounts /articles/oneineightbritishhouseholdshasnogarden/2020-05-14.

US Environmental Protection Agency. "U. S. Environmental Protection Agency's Policy to Mitigate the Acute Risk to Bees from Pesticide Products." Regulations.gov, January 12, 2017. regulations.gov/document/EPA-HQ-OPP-2014-0818-0477.

Van Deynze, Braeden, Scott M. Swinton, David A. Hennessy, Nick M. Haddad, and Leslie Ries. "Insecticides, More Than Herbicides, Land Use, and Climate, Are Associated with Declines in Butterfly Species Richness and Abundance in the American Midwest." *PLoS One* 19, no. 6 (2024): e0304319.

van Kleunen, Mark, Franz Essl, Jan Pergl, et al. "The Changing Role of Ornamental Horticulture in Alien Plant Invasions." *Biological Reviews* 93, no. 3 (2018): 1421–1437.

Wagner, David L., Eliza M. Grames, Matthew L. Forister, May R. Berenbaum, and David Stopak. "Insect Decline in the Anthropocene: Death by a Thousand Cuts." *Proceedings of the National Academy of Sciences* 118, no. 2 (2021): e2023989118.

Way, Twigs. *Virgins, Weeders and Queens: A History of Women in the Garden.* History Press, 2005.

Wong, James. "Gardening with 'Native' Plants Isn't Always Better for the Environment." *New Scientist,* September 27, 2023. newscientist.com/article/mg25934581-200-gardening -with-native-plants-isnt-always-better-for-the-environment.

Yockteng, R., H. E. Ballard Jr., G. Mansion, Isabelle Dajoz, and Sophie Nadot. "Relationships Among Pansies (*Viola* section *Melanium*) Investigated Using ITS and ISSR Markers." *Plant Systematics and Evolution* 241 (2003): 153–170.

Speculative Futures

Acoca-Pidolle, Samson, Perrine Gauthier, Louis Devresse, Antoine Deverge Merdrignac, Virginie Pons, and Pierre-Olivier Cheptou. "Ongoing Convergent Evolution of a Selfing Syndrome Threatens Plant–Pollinator Interactions." *New Phytologist* 242, no. 2 (2024): 717–726.

Akiyama, Reiko, Stefan Milosavljevic, Matthias Leutenegger, and Rie Shimizu-Inatsugi. "Trait-Dependent Resemblance of the Flowering Phenology and Floral Morphology of the Allopolyploid *Cardamine flexuosa* to Those of the Parental Diploids in Natural Habitats." *Journal of Plant Research* 133 (2020): 147–155.

Alberti, Marina. "Cities of the Anthropocene: Urban Sustainability in an Eco-Evolutionary Perspective." *Philosophical Transactions of the Royal Society B* 379, no. 1893 (2024): 20220264.

Argüelles, Lucía, and Hug March. "Weeds in Action: Vegetal Political Ecology of Unwanted Plants." *Progress in Human Geography* 46, no. 1 (2022): 44–66.

Arruda, Andre J., Patricia A. Junqueira, Hanna T. S. Rodrigues, et al. "Limited Seed Dispersability in a Megadiverse OCBIL Grassland." *Biological Journal of the Linnean Society* 133, no. 2 (2021): 499–511.

Baldwin, Maude W., Yasuka Toda, Tomoya Nakagita, et al. "Evolution of Sweet Taste Perception in Hummingbirds by Transformation of the Ancestral Umami Receptor." *Science* 345, no. 6199 (2014): 929–933.

Banaticla-Hilario, Maria Celeste N., Marc S. M. Sosef, Kenneth L. McNally, Nigel Ruaraidh Sackville Hamilton, and Ronald G. van den Berg. "Ecogeographic Variation in the Morphology of Two Asian Wild Rice Species, *Oryza nivara* and *Oryza rufipogon*." *International Journal of Plant Sciences* 174, no. 6 (2013): 896–909.

Born, Julia, H. P. Linder, and P. Desmet. "The Greater Cape Floristic Region." *Journal of Biogeography* 34, no. 1 (2007): 147–162.

Bothe, Hermann, and Aneta Słomka. "Divergent Biology of Facultative Heavy Metal Plants." *Journal of Plant Physiology* 219 (2017): 45–61.

Bradshaw, Peter L., and Richard M. Cowling. "Landscapes, Rock Types, and Climate of the Greater Cape Floristic Region." In *Fynbos: Ecology, Evolution, and Conservation of a Megadiverse Region*, edited by Nicky Allsopp, Jonathan F. Colville, and G. Anthony Verboom. Oxford University Press, 2014.

Brundrett, Mark C. "One Biodiversity Hotspot to Rule Them All: Southwestern Australia—an Extraordinary Evolutionary Centre for Plant Functional and Taxonomic Diversity." *Journal of the Royal Society of Western Australia* 104 (2021): 91–122.

Burd, Martin, C. Tristan Stayton, Mani Shrestha, and Adrian G. Dyer. "Distinctive Convergence in Australian Floral Colours Seen Through the Eyes of Australian Birds." *Proceedings of the Royal Society B: Biological Sciences* 281, no. 1781 (2014): 20132862.

Burton, Rob J. F. "Understanding Farmers' Aesthetic Preference for Tidy Agricultural Landscapes: A Bourdieusian Perspective." *Landscape Research* 37, no. 1 (2012): 51–71.

Chen, Lei, Qiang Qiu, Yu Jiang, et al. "Large-Scale Ruminant Genome Sequencing Provides Insights into Their Evolution and Distinct Traits." *Science* 364, no. 6446 (2019): eaav6202.

Coren, Michael J. "Why 'Chaos Wheat' May Be the Future of Bread." *Washington Post*, September 17, 2024.

Cummings, Claire. "A Conversation with Corn." Lecture at Land Institute Prairie Festival 2002, Salina, Kansas.

Durand-Morat, Alvaro, Lawton Lanier Nalley, and Greg Thoma. "The Implications of Red Rice on Food Security." *Global Food Security* 18 (2018): 62–75.

Fujita, Tomohiro, Naoe Tsuda, Dai Koide, Yuya Fukano, and Tomomi Inoue. "The Flower Does Not Open in the City: Evolution of Plant Reproductive Traits of *Portulaca oleracea* in Urban Populations." *Annals of Botany* 135, no. 1–2 (2024): 269–276.

Gaines, Todd A., Stephen O. Duke, Sarah Morran, et al. "Mechanisms of Evolved Herbicide Resistance." *Journal of Biological Chemistry* 295, no. 30 (2020): 10307–10330.

Goldblatt, P., J. C. Manning, and D. Snijman. "Cape Plants: Corrections and Additions to the Flora. 1." *Bothalia* 35, no. 1 (2005): 35–46.

Gosper, Carl R., David J. Coates, Stephen D. Hopper, Margaret Byrne, and Colin J. Yates. "The Role of Landscape History in the Distribution and Conservation of Threatened Flora in the Southwest Australian Floristic Region." *Biological Journal of the Linnean Society* 133, no. 2 (2021): 394–410.

Hay, Angela S., Bjorn Pieper, Elizabeth Cooke, et al. "*Cardamine hirsuta*: A Versatile Genetic System for Comparative Studies." *Plant Journal* 78, no. 1 (2014): 1–15.

Hensel, Marc J. S., Christopher J. Patrick, Robert J. Orth, et al. "Rise of *Ruppia* in Chesapeake Bay: Climate Change–Driven Turnover of Foundation Species Creates New Threats and Management Opportunities." *Proceedings of the National Academy of Sciences* 120, no. 23 (2023): e2220678120.

Herlihy, Christopher R., and Christopher G. Eckert. "Genetic Cost of Reproductive Assurance in a Self-Fertilizing Plant." *Nature* 416, no. 6878 (2002): 320–323.

Hill, Robert S., and Gregory J. Jordan. "Deep History of Wildfire in Australia." *Australian Journal of Botany* 64, no. 8 (2016): 557–563.

Hopper, Stephen D. "OCBIL Theory: Towards an Integrated Understanding of the Evolution, Ecology and Conservation of Biodiversity on Old, Climatically Buffered, Infertile Landscapes." *Plant and Soil* 322 (2009): 49–86.

Hopper, Stephen D. "Ocbil Theory as a Potential Unifying Framework for Investigating Narrow Endemism in Mediterranean Climate Regions." *Plants* 12, no. 3 (2023): 645.

Hopper, Stephen D., Hans Lambers, Fernando A. O. Silveira, and Peggy L. Fiedler.

"OCBIL Theory Examined: Reassessing Evolution, Ecology and Conservation in the World's Ancient, Climatically Buffered and Infertile Landscapes." *Biological Journal of the Linnean Society* 133, no. 2 (2021): 266–296.

Hu, Wangjie, Ziqian Hao, Pengyuan Du, et al. "Genomic Inference of a Severe Human Bottleneck During the Early to Middle Pleistocene Transition." *Science* 381, no. 6661 (2023): 979–984.

Hultgren, Andrew, Tamma Carleton, Michael Delgado, et al. "Impacts of Climate Change on Global Agriculture Accounting for Adaptation." *Nature* 642, no. 8068 (2025): 644–652.

Judd, Emily J., Jessica E. Tierney, Daniel J. Lunt, et al. "A 485-Million-Year History of Earth's Surface Temperature." *Science* 385, no. 6715 (2024): eadk3705.

Keleman, Alder, Jon Hellin, and M. R. Bellon. "Maize Diversity, Rural Development Policy, and Farmers' Practices: Lessons from Chiapas, Mexico." *Geographical Journal* 175, no. 1 (2009): 52–70.

"Kernza." The Land Institute, 2025. kernza.org.

King Arthur Baking. "Regeneratively-Grown Climate Blend." kingarthurbaking.com /climate-blend.

Knapp, Lynette, Dion Cummings, Shandell Cummings, Peggy L. Fiedler, and Stephen D. Hopper. "A Merningar Bardok Family's Noongar Oral History of Two Peoples Bay Nature Reserve and Surrounds." *Pacific Conservation Biology* 30, no. 3 (2024).

Köhler, Peter, and Roderik S. W. van de Wal. "Interglacials of the Quaternary Defined by Northern Hemispheric Land Ice Distribution Outside of Greenland." *Nature Communications* 11, no. 1 (2020): 5124.

Kruger, F. J., and R. C. Bigalke. "Fire in Fynbos." In *Ecological Effects of Fire in South African Ecosystems. Ecological Studies*, vol. 48, edited by Peter V. Booysen and Neil M. Tainton. Springer, 1984.

Lambers, Hans, Mark C. Brundrett, John A. Raven, and Stephen D. Hopper. "Plant Mineral Nutrition in Ancient Landscapes: High Plant Species Diversity on Infertile Soils Is Linked to Functional Diversity for Nutritional Strategies." *Plant and Soil* 348 (2011): 7–27.

Lee, Kristin M., and Graham Coop. "Distinguishing Among Modes of Convergent Adaptation Using Population Genomic Data." *Genetics* 207, no. 4 (2017): 1591–1619.

Leng, Guoyong, and Jim Hall. "Crop Yield Sensitivity of Global Major Agricultural Countries to Droughts and the Projected Changes in the Future." *Science of the Total Environment* 654 (2019): 811–821.

Louette, Dominique, André Charrier, and Julien Berthaud. "In Situ Conservation of Maize in Mexico: Genetic Diversity and Maize Seed Management in a Traditional Community." *Economic Botany* 51 (1997): 20–38.

Lowder, Sarah K., Marco V. Sánchez, and Raffaele Bertini. "Which Farms Feed the World and Has Farmland Become More Concentrated?" *World Development* 142 (2021): 105455.

Mabry, Makenzie E., Muthukumar V. Bagavathiannan, James M. Bullock, et al. "Building a Feral Future: Open Questions in Crop Ferality." *Plants, People, Planet* 5, no. 5 (2023): 635–649.

Mandáková, Terezie, Judita Zozomová-Lihová, Hiroshi Kudoh, Yunpeng Zhao, Martin A. Lysak, and Karol Marhold. "The Story of Promiscuous Crucifers: Origin and Genome Evolution of an Invasive Species, *Cardamine occulta* (Brassicaceae), and Its Relatives." *Annals of Botany* 124, no. 2 (2019): 209–220.

Maruyama, Pietro Kiyoshi, Camila Bosenbecker, João Custódio F. Cardoso, et al. "Urban Environments Increase Generalization of Hummingbird–Plant Networks Across

Climate Gradients." *Proceedings of the National Academy of Sciences* 121, no. 48 (2024): e2322347121.

Mason, Chase M. "How Old Are Sunflowers? A Molecular Clock Analysis of Key Divergences in the Origin and Diversification of *Helianthus* (Asteraceae)." *International Journal of Plant Sciences* 179, no. 3 (2018): 182–191.

McAlvay, Alex C., Anna DiPaola, A. Catherine D'Andrea, et al. "Cereal Species Mixtures: An Ancient Practice with Potential for Climate Resilience. A Review." *Agronomy for Sustainable Development* 42, no. 5 (2022): 100.

Meng, Qing-Lin, Cheng-Gen Qiang, Ji-Long Li, et al. "Genetic Architecture of Ecological Divergence Between *Oryza rufipogon* and *Oryza nivara*." *Molecular Ecology* 33, no. 5 (2024): e17268.

Montanarella, Luca. "Agricultural Policy: Govern Our Soils." *Nature* 528, no. 7580 (2015): 32–33.

Montanarella, Luca, Daniel Jon Pennock, Neil McKenzie, et al. "World's Soils Are Under Threat." *Soil* 2, no. 1 (2016): 79–82.

Morinaga, Shin-Ichi, Atsushi J. Nagano, Saori Miyazaki, et al. "Ecogenomics of Cleistogamous and Chasmogamous Flowering: Genome-Wide Gene Expression Patterns from Cross-Species Microarray Analysis in *Cardamine kokaiensis* (Brassicaceae)." *Journal of Ecology* 96, no. 5 (2008): 1086–1097.

Mucina, Ladislav, and Grant W. Wardell-Johnson. "Landscape Age and Soil Fertility, Climatic Stability, and Fire Regime Predictability: Beyond the OCBIL Framework." *Plant and Soil* 341 (2011): 1–23.

Mullaney, Emma Gaalaas. "Geopolitical Maize: Peasant Seeds, Everyday Practices, and Food Security in Mexico." *Geopolitics* 19, no. 2 (2014): 406–430.

Mullaney, Emma Gaalaas. "Working with Maize." PhD diss., The Pennsylvania State University, 2021.

Noy-Meir, Imanuel. "Ecology of Wild Emmer Wheat in Mediterranean Grasslands in Galilee." *Israel Journal of Plant Sciences* 49, no. supl (2001): 43–52.

Orians, Gordon H., and Antoni V. Milewski. "Ecology of Australia: The Effects of Nutrient-Poor Soils and Intense Fires." *Biological Reviews* 82, no. 3 (2007): 393–423.

Papadopulos, Alexander S. T., Andrew J. Helmstetter, Owen G. Osborne, et al. "Rapid Parallel Adaptation to Anthropogenic Heavy Metal Pollution." *Molecular Biology and Evolution* 38, no. 9 (2021): 3724–3736.

Prescott, Cindy E. "Sinks for Plant Surplus Carbon Explain Several Ecological Phenomena." *Plant and Soil* 476, no. 1 (2022): 689–698.

Prothero, Donald R. "Species Longevity in North American Fossil Mammals." *Integrative Zoology* 9, no. 4 (2014): 383–393.

Pyke, Graham H., and Nickolas M. Waser. "The Production of Dilute Nectars by Hummingbird and Honeyeater Flowers." *Biotropica* (1981): 260–270.

Robertson, Francesca, Glen Stasiuk, Noel Nannup, and Stephen D. Hopper. "Ngalak Koora Koora Djinang (Looking Back Together): A Nyoongar and Scientific Collaborative History of Ancient Nyoongar Boodja." *Australian Aboriginal Studies* 1 (2016): 40–54.

Robins, Timothy P., Rachel M. Binks, Margaret Byrne, and Stephen D. Hopper. "Contrasting Patterns of Population Divergence on Young and Old Landscapes in *Banksia seminuda* (Proteaceae), with Evidence for Recognition of Subspecies." *Biological Journal of the Linnean Society* 133, no. 2 (2021): 449–463.

Ru, Yalu, Rainer Schulz, and Marcus A. Koch. "Successful Without Sex—the Enigmatic Biology and Evolutionary Origin of Coralroot Bittercress (*Cardamine bulbifera*, Brassicaceae)." *Perspectives in Plant Ecology, Evolution and Systematics* 46 (2020): 125557.

Sánchez González, José de Jesús, José Ariel Ruiz Corral, Guillermo Medina García, et al. "Ecogeography of Teosinte." *PLoS One* 13, no. 2 (2018): e0192676.

Silveira, Fernando A. O., Peggy L. Fiedler, and Stephen D. Hopper. "OCBIL Theory: A New Science for Old Ecosystems." *Biological Journal of the Linnean Society* 133, no. 2 (2021): 251–265.

Smith, Pete, Rosa M. Poch, David A. Lobb, et al. "Status of the World's Soils." *Annual Review of Environment and Resources* 49 (2024).

Southwest Australia Ecoregion Initiative, and Cheryl Gole. *The Southwest Australia Ecoregion: Jewel of the Australian Continent*. Southwest Australia Ecoregion Initiative, 2006.

Stock, W. D., and O. A. M. Lewis. "Soil Nitrogen and the Role of Fire as a Mineralizing Agent in a South African Coastal Fynbos Ecosystem." *Journal of Ecology* 74 (1986): 317–328.

Suijkerbuijk, Hanneke A. C., Sergio E. Ramos, and Erik H. Poelman. "Plasticity in Plant Mating Systems." *Trends in Plant Science* 30, no. 4 (2024): P424–P436.

Timme, Ruth E., Beryl B. Simpson, and C. Randal Linder. "High-Resolution Phylogeny for *Helianthus* (Asteraceae) Using the 18S-26S Ribosomal DNA External Transcribed Spacer." *American Journal of Botany* 94, no. 11 (2007): 1837–1852.

Toda, Yasuka, Meng-Ching Ko, Qiaoyi Liang, et al. "Early Origin of Sweet Perception in the Songbird Radiation." *Science* 373, no. 6551 (2021): 226–231.

Tola, Maya M. "The Allure of Ruins—Architectural Decay in Art." *Daily Art Magazine*, November 8, 2024. dailyartmagazine.com/ruins-in-art.

Tossi, Vanesa E., Leandro J. Martínez Tosar, Leandro E. Laino, et al. "Impact of Polyploidy on Plant Tolerance to Abiotic and Biotic Stresses." *Frontiers in Plant Science* 13 (2022): 869423.

University of Cambridge Department of Earth Sciences. "New Record of Earth's Cenozoic Climate Reveals Defining Role of Polar Ice." September 3, 2020. esc.cam.ac.uk/about-us/news/record-earths-cenozoic-climate-reveals-role-of-polar-ice.

Van de Peer, Yves, Tia-Lynn Ashman, Pamela S. Soltis, and Douglas E. Soltis. "Polyploidy: An Evolutionary and Ecological Force in Stressful Times." *Plant Cell* 33, no. 1 (2021): 11–26.

Wang, Jin, Sai Kranthi Vanga, Rachit Saxena, Valérie Orsat, and Vijaya Raghavan. "Effect of Climate Change on the Yield of Cereal Crops: A Review." *Climate* 6, no. 2 (2018): 41.

Westerhold, T., N. Marwan, A. J. Drury, et al. "An Astronomically Dated Record of Earth's Climate and Its Predictability over the Last 66 Million Years." *Science* 369, no. 6509 (2020): 1383–1387.

Wright, Stephen I., Susan Kalisz, and Tanja Slotte. "Evolutionary Consequences of Self-Fertilization in Plants." *Proceedings of the Royal Society B: Biological Sciences* 280, no. 1760 (2013): 20130133.

Supplement: Invitations to Play with Flowers

Badger, Marc, Victor Manuel Ortega-Jimenez, Lisa von Rabenau, Ashley Smiley, and Robert Dudley. "Electrostatic Charge on Flying Hummingbirds and Its Potential Role in Pollination." *PLoS One* 10, no. 9 (2015): e0138003.

Clarke, Dominic, Heather Whitney, Gregory Sutton, and Daniel Robert. "Detection and Learning of Floral Electric Fields by Bumblebees." *Science* 340, no. 6128 (2013): 66–69.

Clarke, Dominic, Erica Morley, and Daniel Robert. "The Bee, the Flower, and the Electric Field: Electric Ecology and Aerial Electroreception." *Journal of Comparative Physiology A* 203 (2017): 737–748.

England, Sam J., and Daniel Robert. "The Ecology of Electricity and Electroreception." *Biological Reviews* 97, no. 1 (2022): 383–413.

England, Sam J., and Daniel Robert. "Electrostatic Pollination by Butterflies and Moths." *Journal of the Royal Society Interface* 21, no. 216 (2024): 20240156.

Gange, Alan C., and Annabel K. Smith. "Arbuscular Mycorrhizal Fungi Influence Visitation Rates of Pollinating Insects." *Ecological Entomology* 30, no. 5 (2005): 600–606.

Guzman, Aidee, Marisol Montes, Nada Lamie, et al. "Arbuscular Mycorrhizal Interactions and Nutrient Supply Mediate Floral Trait Variation and Pollinator Visitation." *New Phytologist* 245, no. 1 (2025): 406–419.

Jones, Michelle L. "Mineral Nutrient Remobilization During Corolla Senescence in Ethylene-Sensitive and -Insensitive Flowers." *AoB Plants* 5 (2013): plt023.

Khan, Shahmshad Ahmed, Khalid Ali Khan, Stepan Kubik, et al. "Electric Field Detection as Floral Cue in Hoverfly Pollination." *Scientific Reports* 11, no. 1 (2021): 18781.

Montgomery, Clara, Jozsef Vuts, Christine M. Woodcock, et al. "Bumblebee Electric Charge Stimulates Floral Volatile Emissions in *Petunia integrifolia* but Not in *Antirrhinum majus*." *Science of Nature* 108 (2021): 1–12.

Sutton, Gregory P., Dominic Clarke, Erica L. Morley, and Daniel Robert. "Mechanosensory Hairs in Bumblebees (*Bombus terrestris*) Detect Weak Electric Fields." *Proceedings of the National Academy of Sciences* 113, no. 26 (2016): 7261–7265.

Vaknin, Yiftach, Samuel Gan-Mor, Avital Bechar, Beni Ronen, and Dan Eisikowitch. "The Role of Electrostatic Forces in Pollination." In *Pollen and Pollination*, edited by Amots Dafni, Michael Hesse, and Ettore Pacini. Springer, 2000.

Woodburn, F. A., L. J. O'Reilly, L. Bentall, and D. Robert. "Electrostatic Detection and Electric Signalling in Plants: Do Flowers Act as Antennas?" *Journal of Physics: Conference Series* 2702, no. 1 (2024).

Index

abolitionism, 180
"abominable mystery," (Charles Darwin)
    34, 36
Aboriginal Australians, 247, 251
acacias, 133, 247
Adder's-tongue ferns, 41
aesthetics, xv–xvi, 10, 239–40
    floral beauty, 255–57
    pansies and, 189–90
    rewilding, 194, 210–11, 217
Africa
    grasses and grasslands, 77, 79, 96,
        99–101
    monkeys, 150
African Cape Floristic Region,
    250–51
agricultural revolution, 39, 93–94
agriculture, xiii, 27, 79, 88–90, 94, 124,
    234–38
algae, xii, 35, 106, 116, 117, 121, 124
alkaloid poisoning, 93
amaranth, 227
Amborella, 4
American Museum of Natural History,
    xv–xvi
Andrle, Gabe, 74–76, 78–82
angeion, 19
angiosperm, 19

antimicrobials, 151, 154
aphids, 199
Arabia Mountain, 241–43
Archaeanthus, 4–5
Ardipithecus ramidus, 100
Argentina, 85
Argogorytes mystaceus, 59
Arlott, Norman, 57
Arnold, Naomi, 122
aroma, 9–10, 140–56. *See also* perfume
    evolution and, 7–8, 140–42, 146
    funerary aromas, 151–53
    genetics and, 142–43
    human evolution and, 146–47,
        149–50, 153
    humans and culture, 147–49,
        151–53
    magnolias and, 7, 9–10, 15–19
    orchids and, 49–50, 58, 59, 65, 66, 67,
        69–70, 147, 172
    roses and, 132–33, 139–40, 143–44,
        147–48, 151–52
    volatiles, 139–43
aroma receptors, 149–50, 155, 156
arums, 118, 142, 147
Assyrians, 147, 154
asters, 18, 65, 86, 193, 194, 213, 216, 218,
    251, 268

INDEX

Atlanta, Georgia, 3, 6, 21, 75, 81, 84, 89,
    152, 278
    Arabia Mountain, 241–43
Australasian figbirds, 244
Australia, 60, 61, 108, 226, 243–50
Australopithecus africanus, 100
awn, 92
Axel, Richard, 156

bacteria, 9, 86, 91, 247–48
    aromas and, 151, 154
    seagrasses and, 117–18
Balance, The, garden, 204–5, 206
Balearic Islands, 108
bananas, 177–78
barley, 39, 88, 234, 238
"barrels of gold," 174–75, 176, 178
bat-pollinated flowers, 141–42
Beaux, Ernest, 148
bee orchids, 58, 60, 69
bees, 144–46, 281, 282
    aroma and, 144–47, 155
    caffeine and, 63
    chemicals and pesticides, 196, 198
    columbines and, 214
    evolution, 8, 78, 146
    extinction, 211
    in gardens, 193, 194, 205
    genetics, 60
    goatsbeard and, 30
    grasslands and, 84, 87
    orchids and, 60, 62–63, 66, 67, 69–70,
        270, 271
    pansies and, 189, 191
    roses and, 132–33, 139–40, 142
beetles, 60, 64, 87, 142, 147, 191, 195
    magnolias and, 8, 11–13, 15, 17–18
bellflowers, 61
Bennet, Mary Elizabeth, 188–89
Bermuda grass, 75–76, 77
Bernáldez, Andrés, 111
biodiversity, xiii, 113, 208, 251

birds. *See also specific birds*
    aromas and, 141–42, 153
    eelgrass and, 109
    evolution, 8, 248–49
    grasses and, 74–75, 80, 81, 82, 84, 87
    magnolias and, 21–22
    pansies and, 192
Birds Georgia, 74–75, 81–84, 87, 126
bird's-nest orchids, 69
bisexuality, 5, 11, 52, 170, 280
    of magnolias, 5, 11, 14–15, 19, 20
bittercress, 225–28, 231, 233–34, 239–40
black-eyed Susans, 84, 86
Blackness Castle, 105–6, 107, 119
Blackness, Falkirk, 105–7, 109–10,
    112–14
blazing stars, 86
bluebirds, 81
bluestem, 75, 80, 82, 86, 98
bobwhite quail, 80–81
body scent, 155–56
bog stars, 62
bonobos, 150
botanic gardens, xvi–xvii, 178–79.
    See also specific gardens
Boyce, C. Kevin, 43
Boyle, Lyle, 119–27
brain
    aromas and smells, 146–47, 149,
        150, 155
    evolution, 101–2, 256
bromeliads, 68
Bronze Age, 237
Brooklyn, 211–20
Brooklyn Botanic Garden, 214
Brooklyn Bridge, 216, 242
Brooklyn Bridge Park, 215–20
Brooklyn Museum, 212–14
broomsedge, 75, 76–77, 79, 80, 82,
    83–84, 87, 98
Buck, Linda, 156
Buddha, 255

bumblebees, 67, 143–44, 189, 191, 194, 211, 229, 278, 280
bunchgrasses, 77, 80, 86, 133
Burner Bob, 80–81
Burntisland, 119–27
Burpee Seeds, 187
butterflies, 8, 196, 211

C3 plants, 98–99
C4 plants, 98–102
caffeine, 63
calcium, 49
Camellia, 174–75, 178
Campanula, 61
camphorweed, 84
canna lilies, 164–65, 171–72
Cantelo, Reg, 185–87, 189, 199–200
capuchin monkeys, 150
carbon dioxide, 41–43, 96–102, 122, 127–28
carbon sinks, 87, 109, 113, 122
carpenter bees, 66, 132–33
carrion flies, 64
Carthage, sack of, 232
Catalogue of Plants and Seeds (Kennedy and Lee), 187
caterpillars, 143–44, 205, 212
  chemicals and pesticides, 195, 196, 199
  pansies and, 191–92, 193
Catherine de' Medici, 135
centifolia, 133
Chanel No. 5, 148
Charmantier, Isabelle, 161–62, 164–65, 166, 174–75, 177–78
cheeses, 147
Chelsea Flower Show, 201–10, 238
chickpea, 27
chickweed, 227
Childs, Ken, 50
Chiltern Hills, 57, 85
chimpanzees, 100, 150, 153
China, 220

chrysanthemums, 195
  magnolias, 16
  peonies, 190
  perfumery, 138, 153
  tea, 174, 175, 178
  wheat and rice, 238
chlorophyll, 55, 95, 96
chloroplasts, 95, 96, 116
Christchurch Horticultural Society, 185–87
chromosomes, 30–32, 34–35, 39, 40, 41–43, 143
chrysanthemums, xvii, 194–95, 197, 202
cinchona, 163
citrus, 63
clams, 56, 110, 117, 118
clematis, 202
Clifford, George, 162–65, 177–78
Clifford's bananas, 177
Climate Blend flour, 235
climate change, 241, 252
climate crisis, 113, 235, 236, 240, 241
clonal growth, 117
clovers, 86, 191
codependency, 47, 269
Coeur d'Alene, 27
coffee, xvi, 37, 63, 163, 178–79
Cole, Thomas, 239–40
colonialism, xvi, 161, 163, 179
color receptors, 10, 149
columbines, 170, 190, 191, 213, 214
Columbus, Christopher, 111
compost, 19, 131, 207–8
conifers, 4, 14, 17, 42, 68
corals, 56, 112, 117
corn, 91, 94–96, 97–99
cornmeal, 88–89
corpse flowers, 147
Course of Empire (Cole), 239–40
COVID-19 pandemic, 25
crabs, 109–10
cranefly orchids, 48–52, 70

creationism, 169–70

creek sedge, 217

Cretaceous, xii–xiii, 3–4, 7, 42, 43, 76, 233, 243

Cretaceous-Paleogene extinction event, xii–xiii, 39–40, 112

crocuses, 62, 165

Cromwell, Oliver, 107

crossbreeding, 37–38, 187–89

Cryptogamia, 167

cut-flower industry, 196–98

Cuvier, Georges, 171

cycads, 17, 20, 21

Cycnoches, 56

Cyzicus, 148

daisies, 11, 190, 193, 268

Damask roses, 133

dandelions, 25, 172, 191, 206, 209–10. *See also* goatsbeard

Darius III, 154

Darwin, Charles, 33, 34, 36, 47, 56, 70, 170–71, 187–88

    Descent of Man, The, 180–81

    On the Origin of Species, 33, 70, 170–71

Davies, Jon, 204–5

dayflowers, 62

Dekalb Farmers Market, 89

delphiniums, 166, 202

De materia medica (Dioscorides), 154

Descent of Man, The (Darwin), 180–81

Desiderata (Ehrmann), 77

Diandria, 165, 166

Dickinson, Emily, 81

Dinsmore, Addie, 126–27

Dioscorides, 154

Disa pulchra, 61

disease-causing species, 53–54

distillation, 137–38, 273

DNA, 30–40, 66–67, 91, 142–43, 171–72

Dobzhansky, Theodosius, 33

dogfennel, 84

Donne, John, 77

drooping figs, 141

drying flowers, 264–65

dugongs, 110–11

durian fruits, 147

Dutch East India Company, 163–64

Eastern Parkway, Brooklyn, 212–14

eastern phoebes, 81

ecological loneliness, 151

Ecuador, 197

Edison, Thomas, 5

eelgrass, 106–10, 113, 114, 115, 120–24, 127–28

Egyptians, ancient, 89, 148, 153–54

Ehrmann, Max, 77

Eiseley, Loren, 9

ekstasis, 7

Elizabeth I of England and Ireland, 188

Encyclopædia Britannica, 167

endosperm, 90–92

enfleurage, 138–39, 145

Eocene, 127–28

epiphytes, 68–69

Escherichia coli, 154

Ethiopia, 100, 238

eugenics, 180–81

Euglossa, 146

Eurasian robins, 202–3

European Union, 198

evening primrose, 18

evolution, xii–xiii, xiv, 3–9, 34, 43–44, 225

    goatsbeard and, 29–34, 38–39

    grasses and, 73–74, 78–79, 90–94

    of humans. *See* human evolution

    Linnaeus and, 169–73, 180–81

    magnolias and, 4–5, 6–7, 14

    natural selection, 32, 35, 36, 44, 47, 64, 68, 230

    orchids and, 56–57, 66–68, 70

    pansies and, 187–90

seagrasses and, 114–17, 118–19,
123–24, 127–28
weeds and, 229–30
extinction crisis, 113, 211, 239, 252

famines, 88, 175–76
fashion and aroma, 148–49
fecal aromas, 142, 147–48
feminine, flowers as, xi–xii, 15, 147
fermentation, 91, 147
ferns, 4, 36, 41, 42
Fertig, Walter, 28, 29, 30
fertilizers, 55, 88, 118, 237
*Field Guide to the Orchids of Britain and
Europe, A* (Williams), 57
field mustard, 143–44
field pansies, 229
figbirds, 244
figs, 141, 142, 244–46
figwort, 152
fire, 74–78, 243, 249–50
grasses and, 74–78, 80–83
"first flower," 3, 5, 37, 78, 200
Firth of Forth, 105–7, 109–10, 112–14
seagrass restoration, 119–27
Firth of Forth Lobster Hatchery, 120
fish, 108, 110, 112
Fisher, Ronald, 33
flies, 8, 13, 14, 60, 64, 69, 147
pollinated flowers, 141–42
floral aroma. *See* aroma
floral heating. *See* heat
floret, 92
flounders, 110
flour, 38, 88–92, 235–36
flower diary, 276–77
flowers, use of term, xvii
fly orchids, 48–52, 58–60, 67, 69, 70
foamflowers, 218
Folkard, Richard, 154
food rations, 185, 186
forager bees, 144–45

forest canopies, 4, 80, 110
Forest Unseen, The (Haskell), 278
Forster, E. M., 48
fossil flowers, xv, 3–5, 13
Foulon spring, 136–37
Freeman, J., 187
fruits, 8, 9, 20, 150
funerary aromas, 151–53
fungi, 9, 238, 247, 249
chemicals and, 195
DNA, 30
grasses and, 87, 91
magnolias and, 20, 22
orchids and, 47, 48, 51–56, 67, 68, 90
fungicides, 195–96, 197

gamagrass, 86
gardens, 185–221
Brooklyn Bridge Park, 215–20
Brooklyn Museum, 212–14
Chelsea Flower Show, 201–10, 238
chemicals and pesticides, 195–200
Christchurch Horticultural Society,
185–87
of George Clifford, 162–65, 177–78
insects in, 194–95, 205–6, 210–11
International Perfume Museum
Gardens (Grasse), 131–34, 136–37,
139, 143
Jardin du Roi (Paris), 167–68, 178–79
of Mary Elizabeth Bennet, 188–89
peat-free, 206–8
rewilding, 194, 210–11, 217
vegetable, 25, 27, 38, 152, 185–86, 195
"weeds" in, 25, 131, 226–27
Gathaagudu, 108, 115, 122
gene duplication, 35–41, 43–44, 67, 94,
116, 142–43
gene mutation, 35, 40–41, 93, 94, 190,
231–32, 265
genes and genetics, 34–41, 43–44
aromas and, 142–43

genes and genetics (*cont.*)
  eugenics, 180–81
  goatsbeard, 30–34, 38, 233
  grasses, 91, 93–94, 99
  orchids, 66–67, 142
  seagrasses, 116
  weeds, 229
Genesis, 94, 169
geologic time, 33, 34, 243
Georgia Department of Natural Resources, 82
gerberas, 197
ginkgo trees, 11, 20
giraffes and grasses, 100
gladioli, 202
globeflowers, 62
glume, 92
gnats, 63–64, 199
goatsbeard, 25–44, 233–34
  evolution, 29–34, 38–39
  genetics, 30–34, 38, 233
  hybridization, 26, 30–34, 38, 40, 233
  name and taxonomy, 25–26, 28–29, 177
  Ownbey's study of, 26–31, 32
  sexuality, 30–32
gold, 138, 174–75, 176, 178
goldenrods, 86, 193, 194, 216, 278
Goring & Wright, 187
governmental regulation, 198–99
Grangemouth oil refineries, 110
grass apes, 88, 102
Grasse, perfume industry, 133–39, 147, 190
grasses, xiii, 73–102. *See also specific grasses*
  agriculture and food, 8, 88–90, 94–102, 171, 234–38
  C3 and C4, 98–102
  endosperm, 90–92
  evolution, 73–74, 78–79, 90–94
  fire and, 74–78, 80–83
  genetics, 91, 93–94, 99

grazing animals, 78–80, 86–88, 88, 93, 99–100
  leaves, 74–79, 95–96
grasshoppers, 9, 82, 87, 100
grasslands, 39–40, 73–74, 79–80
  C3 and C4, 98–102
  diversity of, 85–87
  Panola Mountain State Park, 74–76, 79, 80, 81–85, 92–93
  use of term, 85
  wildflowers in, 76, 84–86
grass stems, 76–77, 91
grazing animals, 78–80, 86–88, 88, 93, 99–100
Great Dividing Range, 226
Greek sage, 152
Greeks, ancient, xv, 7, 19, 88, 154
green sea turtles, 111–12, 117
Grevillea, 246, 250
Grindley, Julia, 58, 105–7, 109–10, 112–13, 116, 185
Guangzhou, 220, 225
Gustave III of Sweden, 180
Gutenberg, Johannes, 5
Gynandria, 165

Haldane, J.B.S., 33
Haskell, Jean, 57
hatchet clams, 117, 118
hawkmoths, 67, 144
hawks, 82
hawthorn, 69, 141–42
hay fever, 74
head lice, 53–54
heat
  grasses and, 77, 83, 86, 95, 97–100, 102, 238
  pollination and, 16–17
  seagrasses and, 122
hedgerows, 192, 216
Henry VIII of England, 107
herbicides, 76, 84, 231–32

herbivores, 8, 62, 69, 76, 93, 99, 100, 116, 143–44

"hero narratives," 5–6

hippos and grasses, 100

hogs and grasses, 100

Homer the Lobster, 120

Homo erectus, 101

Homo poaceae, 88, 102

Homo sapiens, 100–101

honeybees, 62–63, 132–33, 139–40, 144–45, 149, 191

honeyeaters, 8, 246, 247

Hooker, Joseph, 34

horse manure, 131

horses and grasses, 100

horseweed, 227

Hortus Cliffortianus (Linnaeus), 161–62, 164–65

"hotspots," 251

hoverflies, 67, 69, 70, 84, 131, 193, 205, 281

Hughes, Becky, xi

human evolution, 256
    aromas and, 146–47, 149–50, 153
    grasses and grasslands, 8, 39, 94–102
    March of Progress, 102

humans and body scent, 155–56

hummingbirds, 8, 17, 144, 191, 213–14, 279

Hunt, Adam, 210

hurricanes, 89, 121

hyacinths, 133–34

hybridization, 94, 188–90
    goatsbeard, 26, 30–34, 38, 40, 233

India, 237–38

Indian grass, 83–84

indigenous peoples, 27, 80, 83, 105, 113, 227, 251

indigo, xvi, 84, 86, 163, 170

indigo buntings, 83

indole, 142

industrialization, 113, 122

insecticides, 196, 199

insects, xv. *See also specific insects*
    aromas and, 140–42, 145
    evolution, 8
    in gardens, 194–95, 205–6, 210–11
    grasses and, 86–87
    orchids and, 65–66, 69–70
    pansies and, 189–90, 191–92
    pollination, 11, 13–18
    sitting with flowers and, 279–80

interbeing, 157

interconnection, 47, 55, 126

interdependence, 44, 47, 51–56

International Perfume Museum Gardens, 131–34, 136–37, 139, 143

irises, 193

iron deficiency, 116

ironweeds, 217

Isaiah, 87–88

IUCN "Red List" of Threatened Species, 239

Jack-in-the-pulpits, 63–64

Japan, xvii, 148

Japanese irises, 193

Jardin du Roi (Paris), 167–68, 178–79

jasmine, 131, 133, 136, 137, 138, 147, 148, 150, 154, 274

Johnny-jump-ups, 187

joy, 256–57

Judean sage, 152

"jumping genes," 66–67

Jussieu, Antoine de, 179, 188

Kass, Deborah, 213

Kentucky coffeetrees, 218

Kenya, 85, 101

kestrels, 82

King Arthur Baking Company, 235

kinship, 64, 170–74, 180, 280

Kipling, Rudyard, 169
Klausing, Brook, 212

lady orchids, 58, 69
La Mettrie, Julien Offray de, 166–67
Land Institute, 235–36
larkspur, 93, 190
La Roche-Guyon, 57–61, 65–66, 69
Late Pleistocene die-off, 79
lavender, 133, 196, 275
lawn grasses, 75–78, 194
leafhoppers, 87
leaf miners, 192, 199
Lee, James, 188–89
legumes, 40, 86, 233, 235, 251
Lehman, Katherine "Katie," 25, 89, 152,
    193, 212, 244, 245, 247, 248
lemma, 92
L'homme-plante (La Mettrie), 166–67
lignin, 114, 115–16
lilies, xvii, 4, 11, 61, 154
    canna, 164–65, 171–72
    how to read, 267–68
limonene, 142
linalool, 150
Linnaeus, Carl, 161–81
    coffee, 163, 178–79
    evolution and, 169–73, 180–81
    Hortus Cliffortianus, 161–62, 164–65
    interest in importing plants to
        Sweden, 177–79
    interest in self-sufficiency, 175–77
    legacy of, 180–81
    organizing collections of plants,
        168–69
    sexual system of classification,
        162–68, 170–73, 177, 179–80
    Systema Naturae, 162, 169, 170,
        179–80
    tea, 163, 174–75, 177–78
    two-part classification system, 168
    wealth and privation of, 174–76

Linnean Society of London, 161–67,
    177–78
livestock, 54, 79, 88
lizard orchids, 58, 69
lobelias, 191, 216
lodicule, 92
London, 154, 161, 179
    Chelsea Flower Show, 201–10, 238
    gardens, 194
    World War II Blitz, 161, 186, 253
Long, Edward, 111
lorikeets, 244–46
Louis XIV of France, 135
Louis XV of France, 135, 188

McAlvay, Alex, 238
McMackin, Rebecca, 212–15, 217,
    278–79
magnolias, 3–22
    aroma and, 7, 9–10, 15–19
    birds and, 21–22
    bisexuality, 5, 11, 14–15, 19, 20
    evolution, 4–5, 6–7.
    extinction status, 239
    orchids compared with, 52
    petals, 11–12, 14–18
    pollination, 11–20
maize, xiii, 88, 234, 237–38
malaria, 53–54
manatees, 110–11
mangroves, 108, 109, 112, 122
March of Progress, 102
marigolds, 190, 194–95
market economics, 47, 176
Markfield Park (London), 206
Mary, Queen of Scots, 107
maslins, 237–38
Massey, Tom, 205–6
Maxillaria, 61
Mayr, Ernst, 33
May roses, 132–33, 190
medicines, 26, 153–54

Mège, Christophe, 131–32
Melanesia, 113
meristems, 77
Mesolithic, 108
methyl jasmonate, 150
Metropolitan Museum of Art, xvi
Michael Van Valkenburgh Associates,
	215–16
microbes, xii, 47, 91, 259
Micronesia, 113
Midsummer Night's Dream, A
	(Shakespeare), 58–59
milkweeds, 84, 196, 216
millet, 88, 99
mimicry, 12, 59–64, 67, 69, 142, 147,
	155–56
miner bees, 60, 66
mining, 232–33
mint, 84, 152
mites, 63, 140–41, 199
mollusks, 110
Monandria, 165, 166
monarch butterflies, 196
monkey orchids, 58, 59, 69
monocultures, 85, 194, 205–6, 235, 238
Monogynia, 165
moonshine perfume, 273–75
Moreton Bay figs, 244–46
moths, 8, 49, 50, 67, 210–11
Mount Carmel, Israel, 151–52
movable-type printing press, 5
Musa Cliffortiana, 177
mussels, 110, 174
mustard plants, 67, 88, 143–44
myrcene, 142

narcissus, 133, 137, 142, 154
natural selection, 32, 35, 36, 44, 47, 64,
	68, 230
Neanderthals, 153
nectar spurs, 61, 214
nematodes, 141

Neolithic, 108
neonicotinoids, 198–99
New York Botanical Garden, 238
New York Times, The, xi
Nez Perce, 27
Nietzel, Dietrich, 177
nitrogen, 86, 117, 118, 142, 235,
	247–48
Norton, Jake, 120, 122
"noxious weeds," 227

oil palms, 88
olfactory receptors, 149–50, 155, 156
On the Origin of Species (Darwin), 33, 70,
	170–71
Ophrys, 58, 60–61, 65–66, 67, 69. *See also*
	orchids
opium, 163
orchids, 47–70
	allegories into "nature," 64–65
	aroma and, 49–50, 58, 59, 65, 66, 67,
		69–70, 147
	at Chelsea Flower Show, 202
	"DIY orchid flasking," 272
	evolution, 48, 56–57, 66–68, 70, 90
	extinction status, 239
	fungi and, 47, 48, 51–56, 67, 68, 90
	genetics, 66–67, 142
	greenhouse-grown, 55–56, 270–71
	how to read, 268
	of La Roche-Guyon, 57–61, 65–66, 69
	mimicry, 61–64, 67, 69
	parasitism, 48, 53–55, 61, 63, 70
	pollination, 48–53, 59–60, 61, 65–66,
		69–70, 146, 269–72
	sexual deception, 60–65, 67, 141
	taxonomy, 65, 66, 165
Orkney Islands, 120
ornamental trees, 195
Orwell's Roses (Solnit), 197–98
Ownbey, Marion, 26–31, 32
oxygen, 78, 97, 116, 119

oxygen revolution, xii
oysters, 110, 121, 125, 126

Palau, 113
palea, 92
Paleolithic cave paintings, xiv
Palmer's amaranth, 227
Palouse, 27, 33
Panola Mountain State Park, 74–76, 79, 80, 81–85, 92–93, 126
pansies, 185–93, 229
   crossbreeding, 187–89
   evolution, 187–90
   pollination, 189–90, 191
   transcontinental transplants, 191–93
Paranthropus boisei, 100
Paranthropus robustus, 100
parasitism, 48, 53–55, 61, 63, 70
parfum de mal au dos, 137
parrots, 246–47
pea flowers, 61
pearl seeding, 174–75
peat-free gardening, 206–8
Pentland Hills, 105
perfume, 9, 147–57
   body scent and, 155–56
   culture and fashion, 147–49, 151–55
   games in duty-free or department stores, 275–76
   Grasse's perfume industry, 133–39
   International Perfume Museum Gardens, 131–34, 136–37, 139, 143
   moonshine, 273–75
   origin of term, 154
   scented unguents, 274–75
perfumer bees, 145–46, 155
periwinkles, 110, 227
pesticides, 144, 195–200, 207, 231–32
petunias, 18, 190
Phalaenopsis, 55–56, 270–71
Philodendron, 17
photosynthesis, 39, 41–42, 96–102, 116

Pier 6 Uplands, 219
pine cones, 20
pine trees, 11, 81–82
pistils, 164, 165, 166
Place de la Foux (Grasse), 136
plant aromas. *See* aroma
planting local flowers, 263–64
plant metaphors, xiv–xv
plastic flowers, 195
play, 233–34, 263–65
Pliny, 148, 190
pollination, xiii
   aromas and, 140–56
   goatsbeard, 30–31
   grasses, 90–91
   magnolias, 11–20
   orchids, 48–53, 59–60, 61, 65–66, 69–70, 146, 269–72
   roses, 139–40
pollock, 110
pollution, 109–10, 111, 121, 124
poppies, 85, 203, 206, 253
Posidonia australis, 108
Posidonia oceanica, 108, 118
Poulson, Rashid, 215–20
prairies, 27, 39, 80, 85–86, 126, 235
Princeton Conference of 1947, 33
productivity, xiii, 47, 86–89, 112, 193
Project Seagrass, 119
Prospect Park (Brooklyn), 214
Protea, 250–51
pyramidal orchids, 69
pyrodiversity, 74–76, 78

Queen Elizabeth Olympic Park (London), 206
Quercus alba, 168
Quercus rubra, 168
Quesnay, François, 188

racism, 179–81
ragworm, 110

rain (rainfall), 80, 99–100, 241, 250
rainforests, 43, 68–69, 85, 150, 250, 251
rapeseed, 229
razor clams, 110
reciprocity, 44, 48, 64, 181, 200
Red Sea, 111–12
reed canary grasses, 86
Restoration Forth, 119–27
revolution, xii–xiii, 4–5, 43, 96
rewilding, 194, 210–11, 217
rhizosphere, 118
ribbon weeds, 115
rice, 20, 56, 88, 89, 91, 98–99, 99, 238
Richard, William, 188–89
Rift Valley, 101
Robert, Hubert, 239
robins, 81, 217
Rocky Mountains, 85
Romans, ancient, 154, 190, 232
Roosevelt, Theodore, xvi
Rose de Mai, 132–33
roses, 131–57
    aroma and, 132–33, 139–40, 143–44,
        147–48, 151–52
    cut-flower industry, 196–98
    Grasse's perfume industry, 133–39
    how to read, 265–67
    at International Perfume Museum
        Gardens, 131–34, 136–37,
        139, 143
    petals, 138, 139–40, 265–67
    pollination, 132–33, 139–43
rose water, 138–39, 147–48, 154,
    274, 275
Roviana Lagoon, 113
Royal Botanic Gardens, Kew, xvi–xvii
Royal Entomological Society, 205–6
Royal Horticultural Society (RHS),
    201–2, 207–8, 209
Royal Hospital Chelsea, 202–3
rubisco, 96–97, 98
rye, 86, 88, 176, 234

sages, 152
Saillard, Christine, 132
Saint Barthélemy, 180
saliva, 93
saltmarsh asters, 218
salt marshes, 108, 109, 112, 122, 218
salvias, 191
scale insects, 199
scent extraction, 138–39
scientific racism, 179–81
Scottish Seabird Centre (East
    Lothian), 119
Scott, Walter, 67
sea campion, 232–33
seagrasses, 105–28
    carbon and, 109, 113, 122
    evolution, 114–17, 118–19, 123–24,
        127–28
    Firth of Forth, 105–7, 109–10, 112–14,
        119–27
    flowers, 105, 106, 114–15
    grazing, 111–12
    leaves, 107, 108–9, 115–16
    mutualism, 117–18
    reproduction, 115–16
    restoration, 119–27
    sediment gathering, 107–8
seagrass meadows, 107–11, 112, 117, 122,
    240–41
sea level rise, 128, 221, 240–41
sea turtles, 111–12, 117
seaweed, 106, 112, 114, 174
seed catalogs, 187
seeds, 19–21, 230–31, 237
    bittercress, 228
    goatsbeard, 25, 38
    grass, 8, 73–74, 89–94
    orchid, 51–54, 56–57
    pansy, 187, 188, 190, 229
    planting local flowers, 263–64
    sea campion, 232–33
    seagrass, 110, 113, 120–21, 123–27

self-fertilization, 12, 50, 166, 191,
    230–31
sepals, 268, 269, 270
  lilies, 267–68
  magnolias, 18
  roses, 265–67
sexual deception, 12–13
  in orchids, 60–65, 67, 141
sexuality, 229–30. *See also* bisexuality;
    self-fertilization; unisexuality
  goatsbeard, 30–32
  Linnaeus's classification system,
    162–68, 170–73, 177, 179–80
  magnolias, 5, 11–20
  orchids, 52, 60–65, 141, 268, 269–70
  roses, 265–67
  seagrasses, 115–16
  sunflowers, 268–69
sexual segregation, 11
Shakespeare, William, 58–59, 188, 197
Shark Bay (Gathaagudu), 108, 115, 122
"shattering," 93–94
Shechem, 232
shipwrecks, 108
shrubs
  chemicals and, 196
  fire and, 77
Siagne River, 136
Sibbald, Robert, 107
Siegesbeck, Johann Georg, 166–67
silicate, 79, 152
Simpson, George Gaylor, 33
Sinclair, J., 187
sirenians, 110–11
sitting with a flower, 277–82
slavery, xiv, 163, 179, 180–81, 188
slugs, 144, 195
smell. *See* aroma; perfume
Smellie, William, 167
smell receptors, 149–50, 155, 156
Smith, Adam, 47, 176
Smith, James Edward, 161

Smokey Bear, 80
snapdragons, 269
sodium, 116
soil, 9, 195, 247, 251–52
  erosion, 108, 121–22, 241
  goatsbeard and, 26
  grasses and, 86–87, 94, 234–35
  seagrasses, 107–8
Solnit, Rebecca, 197–98
Solomon Islands, 113
songbirds, 191, 216, 248
Sorghastrum nutans, 83–84
sorghum, 88, 99, 237–38
Sounds Wild and Broken (Haskell), 9
soursop, 17
South America, 56, 61, 79
southern magnolias. *See* magnolias
South River, 84
sowing grass, 83–84
Species Plantarum (Linnaeus), 169
spicebush shrubs, 21
spider orchids, 58, 59, 60, 69
spikelets, 93
squashes, 62, 185
stamens, 11, 12, 165, 190
stasis, 7
stigmas, 11, 13, 162–63, 213, 269,
    279, 282
  grasses, 73, 90
  lilies, 267–68
  orchids, 50, 59, 62, 271, 272
  roses, 266–67
  seagrasses, 115
  sunflowers, 268
Stravinsky, Igor, 36
strawberries, 30, 143
stress, 39
stromatolites, 108
sugarcane, xiii, 37, 88, 99, 163
sulfur, 117–18
sumac, 217
Sumerians, 138

sunbirds, 8, 248, 251

sunflowers, 4, 34, 37, 48, 65, 193, 216, 242, 268–69

"superblooms," 85

superweeds, 232–33

sustainability, 204–5, 206

Swedish Academy of Sciences, 178

Swedish East India Company, 178

sweet flag, 133

sweet peas, 202

symbiosis, xiii, 9, 118

symbolism, xi–xii, xiv–xv, 56–57

genetic pruning, 35, 41

synthetics, 133–34

Syria, 138–39

Systema Naturae (Linnaeus), 162, 169, 170, 179–80

tannins, 93

Tanzania, 85

taxonomy, xiv, 34, 161–74, 179–80, 181

tea, 153, 163, 174–75, 177–78

tepals, 18

Thatcher, Margaret, 47

Thomsen, Esther, 120

Thoreau, Henry David, 47, 217

thrips, 13, 199

tongue orchids, 60

Tragopogon, 26–27, 32

Tragopogon mirus, 29, 32

Tragopogon miscellus, 29, 32

trees

    chemicals and, 196

    fire and, 77

Triandria, 165

triple-knot magnolias, 6–7, 11–13, 14–15, 17–22

tulip bulbs, 175–76

tuliptrees, 5

twinflowers, 175

Tyrannosaurus, xv, 3

Uganda, 85

unguents, 139, 153, 274

unisexuality, 14, 163, 170, 269

United Nations Educational, Scientific and Cultural Organization (UNESCO), 134–35

United Nations Environment Programme, 112

Urquhart, Lulu, 210

vegetable gardens, 25, 27, 33, 152, 185–86, 195

Verne, Jules, 95

vetiver grasses, 133

Vikings, 73–74, 91, 94–96

Viola tricolor, 187

violets, 15, 131, 133, 137, 147, 187, 189, 230, 233, 269

Virginia wild rye, 86

Vivaldi, Antonio, 36

volatiles, 139–43, 148, 151, 153, 156

voyeurism, 10

Wadden Sea, 123–24

Washington Square Park (New York City), 278

Washington State University, 26–33, 235

wasps, 8, 58–61, 66, 69–70, 84, 87, 141–42, 246, 279

water lilies, 4, 17

"weed control," 234

weeds. *See also* goatsbeard; *and specific weeds*

    evolution, 229–30

    superweeds, 232–33

    use of term, 226–27

Western Ghats, 85

West Side Story (musical), 26

wheat, 94–96, 98–99

    agriculture production and consumption, 27, 38, 56, 88, 91, 94–96, 98–99, 136–37, 234, 238

wheat (*cont.*)
    evolution, 91, 171, 234
    flour, 88–92, 235–36
    starch, 20, 90, 91
wheat germ, 92
whiteflies, 199
whole wheat flour, 88, 89, 92
Wild City Studio, 204–5
wildflowers, 21, 49, 63, 76, 84–86, 190, 216, 242, 251
wild leeks, 175–76
wild orchids, 18
wild pansies, 187–90, 229
wild rice, 88, 234
wild roses, 133, 143, 265–66
Williams, John and Andrew, 57

Williams, Steve, 204–5
woodpeckers, 21–22, 168, 217
World War I, 185
World War II, 161, 186, 253
World Wildlife Fund, 121, 122, 127
Wright, Sewall, 33

Xenophanes, 154
Xerces Society, 278

yellow pea flowers, 61

zebras and grasses, 100
zoic, xv
zôion, xv
Zostera marina. See eelgrass

About the Author

David George Haskell is a writer and biologist. Known for his integration of science, lyrical writing and close observation of the living world, he has twice been a finalist for the Pulitzer Prize in General Nonfiction, for *The Forest Unseen* and *Sounds Wild and Broken*. In 2024, the American Academy of Arts and Letters granted him an Award in Literature. He is Adjunct Professor of Environmental Sciences at Emory University, a Fellow of the Linnean Society of London and a Guggenheim Fellow.